The
Plastic
Methods
of
Structural
Analysis

The
Plastic
Methods
of
Structural
Analysis

B. G. NEAL

Emeritus Professor, University of London
Formerly Professor of Engineering Structures and
Head of Department of Civil Engineering,
Imperial College of Science and Technology

THIRD EDITION

LONDON NEW YORK
CHAPMAN AND HALL

First published 1956
Reprinted once
Second edition 1963
Reprinted twice
First issued as a Science Paperback 1965
Third (S.I.) edition 1977
Reprinted 1981, 1985

Published by Chapman and Hall Ltd
11 New Fetter Lane, London EC4P 4EE
Published in the USA by Chapman and Hall,
29 West 35th Street, New York NY 10001

ISBN 0 412 21450 4 (Science Paperback edition)

Printed in Great Britain by
Richard Clay (The Chaucer Press) Ltd, Bungay, Suffolk

British Library Cataloguing in Publication Data

Neal, B.G.
    The plastic methods of structural analysis. –
    3rd Ed. – (Science paperbacks)
    1. Plastic analysis (Theory of structures)
    I. Title
    624.1'71   TA652
    ISBN 0-412-21450-4

Library of Congress Cataloging in Publication Data

Neal, B. G. (Bernard George), 1922–
    The plastic methods of structural analysis.

    (Science paperbacks; 13)
    Includes bibliographies and index.
    1. Plastic analysis (Theory of structures)   I. Title.
II. Series.
TA652.N43   1985   624.1'71   85-14998
ISBN 0-412-21450-4 (pbk.)

Contents

Preface

Since the first edition of this book was published in 1956, there has been a widespread acceptance of the concept of limit state design. It is also generally recognized that the appropriate ultimate limit state for many steel frames is plastic collapse, so that the design of such structures is based upon an assessment of the plastic collapse load, an appropriate load factor being provided. Whereas in 1956 the case for the use of the plastic methods had to be argued, this is no longer necessary, and the presentation has accordingly been shortened. The Principle of Virtual Work has been used throughout to unify the treatment.

The book is concerned with the plastic methods of analysis for beams and plane frames, which are based upon the simplifying plastic hinge assumption. It does not discuss the conditions under which members which have entered the plastic range fail by instability. Nor does it deal with other problems of importance in design, such as the behaviour of full-strength welded joints. Nevertheless, the plastic methods, as presented here, can fairly be claimed to be an essential weapon in the armoury of any competent structural designer.

Digital computers are now used extensively to solve structural problems, both of analyis and of design. Some programs which have been developed for frames analyse their behaviour when the simplifying assumptions of the plastic methods are discarded, so that the actual properties of the members are taken into account. Others deal with the optimisation of designs subject to various forms of constraint. These developments have not been dealt with in this edition, although a few are referred to in passing. Only those techniques which are suitable for hand calculation are included; these need to be thoroughly understood as a prelude to the use of computer programs.

Earlier editions of the book contained a comprehensive bibliography. This would now be inappropriate in view of the exclusion of a full discussion of computer-based developments, and so there are few references to the important work of this nature published recently. A selection of references to the classical work which established the basic theory has been retained.

The author is most grateful to Mr John Cima for his excellent work in preparing the illustrations, and to Mrs Eileen Wyatt whose capacity for the speedy production of an accurate typescript is unsurpassed.

London June 1977   B. G. NEAL

The
Plastic
Methods
of
Structural
Analysis

1 Basic Hypotheses

1.1 Plastic hinge and plastic collapse concepts

The plastic methods of structural analysis are now widely used in the design of steel frames, which carry load by virtue of the resistance of their members to bending action. Multistorey, multibay rectangular frames and single or multibay pitched-roof portals are familiar examples of this type of structure, and the definition also includes simply supported and continuous beams. For such structures Baker (1949) pointed out that the most economical and rational designs are achieved by the use of the plastic methods. The plastic methods also have the advantage of simplicity.

The objective of the plastic methods is to predict the loads at which a framed structure will fail by the development of excessive deflections. It is appropriate to begin by examining the behaviour of the simplest type of structure in this category, a simply supported beam carrying a central concentrated load. Fig. 1.1 shows the results of an early test carried out by Maier-Leibnitz (1929) on an I-beam spanning 1.6 m. The beam remained elastic up to a load W of about 130 kN, when the yield stress was attained in the most highly stressed fibres beneath the load. At a load of about 150 kN, the central deflection δ began to increase very sharply for small increases in the load. The beam eventually failed catastrophically by buckling at a load of 166 kN, but before then collapse had already effectively occurred due to the development of unacceptably large deflections.

A slight idealization of the behaviour would be to assume that the deflection could grow indefinitely under a *constant* load of 150 kN, as shown by the broken line in the figure. This assumption disregards the small additional load-carrying capacity which the beam actually possesses above this load, and is therefore conservative. The assumed indefinite growth of deflection under constant load is termed *plastic collapse*, and the load 150 kN at which it occurs is the *plastic collapse load*, denoted by W_c.

This behaviour can be described on the hypothesis that a *plastic hinge* develops at the centre of the beam at the load W_c, when the central bending moment is $0.4\ W_c = 60$ kN m. The characteristic of this hinge is that it can only undergo rotation when the bending moment is 60 kN m, but while the bending moment has this value the rotation can increase indefinitely, thus permitting an indefinite growth of deflection. The bending moment required to develop a

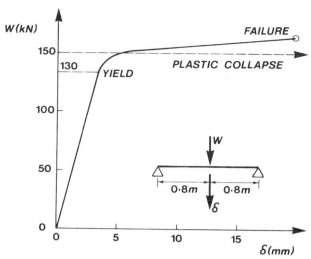

Fig. 1.1 *Test on simply supported beam (after Maier-Leibnitz)*

plastic hinge in this test, 60 kN m, is termed the *plastic moment* of the beam, and is denoted by M_p. It is related to the yield stress of the material, as will be shown in Section 1.3. The plastic methods of analysis, based on the plastic hinge assumption, enable the plastic collapse loads of quite complex frames to be found rapidly, as will be seen in Chapters 3 and 4. Their usefulness as a tool for designing steel frames depends on the fact that large deflections are unlikely to develop before the plastic collapse load is attained. However, it may be necessary to ensure that the deflections developed before collapse are acceptable, and methods for estimating these deflections are discussed in Chapter 5.

The plastic methods should only be used for design if the avoidance of plastic collapse is the governing design criterion. There will be cases in which the primary problem is to avoid other types of failure, for example by fatigue or brittle fracture. These are outside the scope of the simple plastic theory.

It is implicitly assumed throughout that no part of the structure will fail by buckling before the plastic collapse load is reached. The problems of buckling of columns under the conditions actually arising in rigid frames when the members have partially yielded, and of lateral instability and other forms of buckling under similar conditions, have been studied extensively. The pioneering work of J. F. Baker and his associates at Cambridge was presented in *The Steel Skeleton*, vol. 2(1956), and investigations carried out under the direction of Beedle at Lehigh were described in *Plastic Design in Steel* (1971). The present position has been summarized by Horne (1972) and Wood (1972). Rules are available which enable frames to be designed so that failure by certain types of buckling will not occur before the plastic collapse load is attained, but their discussion is outside the scope of this book.

1.2 Stress-strain relation for mild steel

The plastic moment of a steel beam is directly related to the yield stress, as already stated. As a preliminary, it is necessary to review the stress-strain properties of mild steel, the material which is commonly used in the construction of frames.

The relation between direct stress σ and axial strain ϵ for a specimen of annealed mild steel in tension has the typical form shown in Fig. 1.2(a). The relation is linear in the elastic range until the upper yield stress is reached at a. The stress then drops abruptly to the lower yield stress, and the strain then increases at constant stress up to the point b, this behaviour being termed *purely plastic flow*. Beyond b further increases of stress are required to produce further strain increases, and the material is said to be in the *strain-hardening range*. Eventually a maximum stress is reached at c, beyond which the stress decreases due to the formation of a neck in the specimen until rupture occurs at d. The maximum stress is of the order of $400 \, \text{N/mm}^2$ and the strain at fracture is of the order of 0.5.

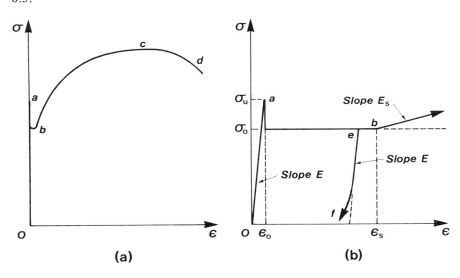

Fig. 1.2 *Stress-strain relation for mild steel in tension*
    (a) Behaviour up to rupture
    (b) Yield range

The yield range Oab is of the most interest from the point of view of plastic theory. Since the strain at b is generally of the order of 0.01–0.02, the yield range can be examined more conveniently if the strain scale is enlarged, as in Fig. 1.2(b). In this figure the upper and lower yield stresses are defined as σ_u and σ_o, respectively, the slope of the initial elastic line Oa is Young's modulus E, and the slope of the initial portion of the strain-hardening line beyond b is defined as E_s. The strains at the yield point a and at the onset of strain hardening b are defined as ϵ_0 and ϵ_s, respectively. If the stress is reduced after yield a relation such as ef is observed, the initial slope being Young's modulus. The deviation from linearity in such an unloading relation is associated with the Bauschinger (1886) effect.

If the stress is increased again after a reduction of this sort, yield occurs at the lower yield stress along eb. This indicates the effect of cold-working in destroying the upper yield stress, which only reappears after further heat treatment.

The values of the constants defined in Fig. 1.2(b) depend markedly on the composition of the steel and its heat treatment, except for the value of Young's modulus, which shows very little variation. Data derived by Roderick and Heyman (1951) from the results of bending tests on four annealed steels of different carbon content are as shown in Table 1.1.

Table 1.1 *Effect of carbon content on properties of steel*

% C	σ_0 (N/mm^2)	$\dfrac{\sigma_u}{\sigma_0}$	$\dfrac{\epsilon_s}{\epsilon_0}$	$\dfrac{E_s}{E}$
0.28	340	1.33	9.2	0.037
0.49	386	1.28	3.7	0.058
0.74	448	1.19	1.9	0.070
0.89	525	1.04	1.5	0.098

It will be seen that the effect of increasing the carbon content is to increase the lower yield stress σ_0 while decreasing the ductility as measured by the ratio ϵ_s/ϵ_0. For structural steel ϵ_s is of the order $10\,\epsilon_0$, and E_s is of the order $0.04\,E$, so that the stress-strain relation is very flat after yield.

It is difficult to determine the actual tensile stress-strain relation of mild steel in the elastic range near the yield point, because of unavoidable eccentricities of loading which cause significant bending stresses. However, Morrison (1939) showed that the initial departure from linearity usually observed below the yield point could be ascribed to yielding in the most highly stressed fibres caused by the eccentricity of loading. He therefore concluded that the yield point, proportional limit and elastic limit were all coincident. The tests also showed that the values of the upper yield stress showed no more variation from specimen to specimen of the same material than those of the lower yield stress. The unpredictable variations in the values of the upper yield stress reported by other observers were therefore concluded to be due to variations in the eccentricity of loading. It was also shown that for a given steel the stress-strain relation in compression is practically identical with that for tension up to the point b where strain-hardening begins.

The yield phenomenon for mild steel is accompanied by the formation of Lüders' lines making an angle of about 45° with the axis of the tensile specimen, showing that plastic flow occurs on those planes where the shear stress is greatest. The material within the Lüders' lines has undergone a considerable amount of slip, corresponding to a jump in the strain from a to b in Fig. 1.2(b). The longitudinal strain in a yielded fibre therefore varies discontinuously along the fibre, and a stress-strain relation such as that shown in Fig. 1.2(b) only represents average strains over a finite length.

The stress-strain relation is often idealized by the neglect of strain-hardening and the Bauschinger effect on unloading, leading to the relation shown in Fig. 1.3(a). Although the upper yield effect is a very real one, it disappears on cold-working and is usually not exhibited by the material of rolled steel sections. Moreover, it will be seen in Section 1.3 that it has no effect on the value of the plastic moment. If it is disregarded, the stress-strain relation becomes that of Fig. 1.3(b), which is often termed the *ideal plastic relation.*

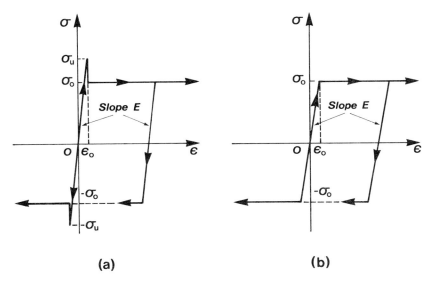

Fig. 1.3 *Stress-strain relations neglecting strain-hardening*

(a) With upper yield stress
(b) Without upper yield stress (ideal plastic)

The neglect of strain-hardening in these idealized relations may seem difficult to justify in view of the fact that the strains will certainly enter the strain-hardening range in many members in actual structures. However, by neglecting the increase of stress during strain-hardening, errors will be introduced which are on the safe side, and it will be seen in Chapter 5 that these errors are usually very small.

1.3 Elastic-plastic bending

For a homogeneous beam of given cross section, the relationship between bending moment and curvature beyond the elastic limit can be derived from the stress-strain relation provided that the usual assumptions of the Bernoulli-Euler theory of bending are made. These are:

(a) The beam is bent by pure terminal couples, so that shear and axial forces are not present.

(b) The deformations are small, so that stresses other than longitudinal normal stresses are negligible.

(c) The relation between longitudinal stress and strain is the same in flexure as in simple tension or compression.

(d) Originally plane cross sections remain plane.

In addition it will be assumed that the stress-strain relation is of the ideal plastic type shown in Fig. 1.3(b), with no upper yield stress. It is further assumed that this relation is obeyed by each individual longitudinal fibre of the beam. In view of the discontinuous nature of the yielding process, this assumption requires experimental verification; several investigators, notably Roderick and Phillipps (1949) have provided evidence in its favour. Finally, it is assumed that there are no residual stresses in the beam. The analysis is simplified considerably if the cross section is symmetrical with respect to an axis which lies in the plane of bending, as happens in many practical cases.

Suppose that the beam is initially straight, and is then bent into an arc of a circle of radius R by pure terminal couples M, say. It is shown in elementary texts on the Strength of Materials that the longitudinal strain ϵ at a distance y from a neutral axis is given by

$$\epsilon = \kappa y \qquad (1.1)$$

where $\kappa = 1/R$ is the curvature of the beam. This relation is derived from purely geometrical considerations, and is independent of the properties of the material. If the beam is initially curved, Equation (1.1) is still true provided that κ denotes the change of curvature produced by M.

1.3.1 *Rectangular cross section*

Consider the rectangular cross section of breadth B and depth D which is shown in Fig. 1.4(a), with the bending moment M acting about an axis Ox parallel to the sides of breadth B. In this case the neutral axis will bisect the cross section, because of its double symmetry.

The linear variation of strain across the section implied by Equation (1.1) is shown in Fig. 1.4(b). Here it is supposed that the strain in the outermost fibres exceeds the strain ϵ_0 which corresponds to the yield stress σ_0 (Fig. 1.3(b)). The yield strain ϵ_0 is attained at distances $\pm z$ from the neutral axis. The corresponding distribution of normal stress is shown in Fig. 1.4(c). There is an elastic core of depth $2z$ outside which there are two yielded zones in which the normal stress is of magnitude σ_0.

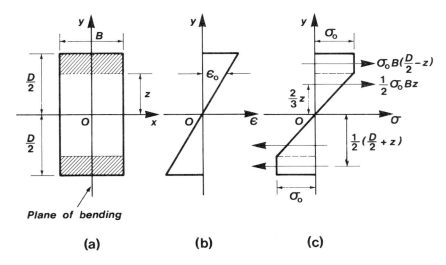

Fig. 1.4 *Elastic-plastic flexure of beam of rectangular cross section*
    (a) Cross section
    (b) Distribution of strain
    (c) Distribution of stress

The bending moment M corresponding to this distribution of stress is readily evaluated. Fig. 1.4(c) shows the resultant normal forces in the two halves of the elastic core and in the yielded zones, and also defines their lines of action. It follows that

$$M = \left(\frac{1}{2}\sigma_0 Bz\right)\frac{4}{3}z + \sigma_0 B\left(\frac{D}{2} - z\right)\left(\frac{D}{2} + z\right) = B\left[\frac{D^2}{4} - \frac{1}{3}z^2\right]\sigma_0.$$

(1.2)

The corresponding curvature is obtained from Equation (1.1) by noting that $\epsilon = \epsilon_0$ when $y = z$, so that

$$\kappa = \epsilon_0/z. \tag{1.3}$$

When $z = D/2$, the yielded zones vanish and the stress only just attains the yield value σ_0 in the outermost fibres. The corresponding bending moment M_y is the greatest moment that the section can withstand before yielding. It is termed the *yield moment*; its value is found from Equation (1.2), with $z = D/2$, to be

$$M_y = \tfrac{1}{6}BD^2\sigma_0. \tag{1.4}$$

M_y could also be found directly from the elastic theory of bending, for by definition

$$M_y = Z\sigma_0, \tag{1.5}$$

where Z is the elastic section modulus, which for a rectangular cross section has the value $BD^2/6$.

The curvature corresponding to this situation is denoted by κ_y, and from Equation (1.3)

$$\kappa_y = 2\epsilon_0/D. \tag{1.6}$$

Combining Equations (1.2)–(1.6), the bending moment-curvature relation can be put in non-dimensional form as follows:

$$\frac{M}{M_y} = 1.5 - 0.5\left(\frac{\kappa_y}{\kappa}\right)^2, \tag{1.7}$$

a result first obtained by de Saint-Venant (1871).

Fig. 1.5 shows this bending moment-curvature relation, together with the elastic relation appropriate when M is less than M_y. Its important feature is that M tends to a limiting value 1.5 M_y as κ becomes very large. In the limit, when $M = 1.5 M_y$, κ becomes infinite, and Equation (1.3) shows that z is then zero, so that the elastic core vanishes. The entire cross section is then plastic, and the corresponding bending moment is the plastic moment M_p. Using Equation (1.4),

$$M_p = 1.5 M_y = \tfrac{1}{4}BD^2\sigma_0. \tag{1.8}$$

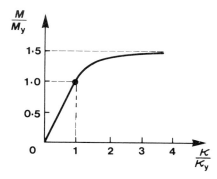

Fig. 1.5 *Bending moment-curvature relation for beam of rectangular cross section*

The attainment of the plastic moment thus corresponds to the development of infinite curvature, which implies that a finite change of slope can occur over an infinitely short length of the beam. This is the explanation of the plastic hinge behaviour observed in steel beams. The stress distributions corresponding to M_y and M_p are shown in Fig. 1.6.

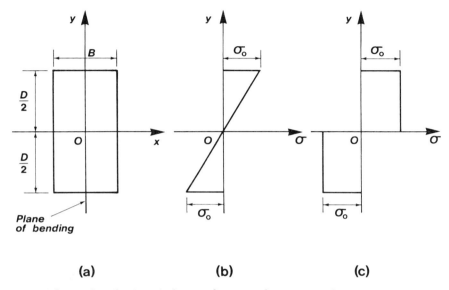

(a)   **(b)**   **(c)**

Fig. 1.6 *Stress distributions in beam of rectangular cross section*

    (a) Cross section
    (b) At yield moment
    (c) At plastic moment

In practice, the condition of full plasticity shown in Fig. 1.6(c) cannot be attained. Equation (1.1) shows that infinite curvature requires infinite strain, which is unattainable. Above a certain curvature, the strains in the outer fibres would become sufficiently large to cause strain-hardening. Suppose, for example, that strain-hardening commences when $\epsilon = 10\,\epsilon_0$. From Fig. 1.4(b), this strain is reached in the outermost fibres when $z = 0.1\,D/2$, and it follows from Equations (1.3) and (1.6) that $\kappa = 10\,\kappa_y$. From Equation (1.7),

$$M = 1.495\,M_y = 0.997\,M_p$$

Thus the bending moment approaches M_p to within 0.3 per cent before strain-hardening begins. The plastic moment can therefore be regarded as an approximate indication of the bending moment at which something very much like a hinge action will occur in practice, with large curvatures developing at the cross section where this moment is attained.

This simple theory assumes that the elastic-plastic boundaries are straight lines parallel to the sides of length B; this is not strictly true, as pointed out by

Hill (1950). Moreover, at large curvatures additional radial stresses would be called into play, for a curved fibre subjected only to tensile or compressive forces at its ends would not be in radial equilibrium. Nevertheless, the plastic moment which corresponds to the fully plastic stress distribution is found in practice to be close enough for all practical purposes to the bending moment which causes plastic hinge action.

1.3.2 Section with a single axis of symmetry

Consider now a beam whose cross section only has one axis of symmetry, as shown in Fig. 1.7(a). O is the centroid of the cross section and Oy is the axis of symmetry, and it is assumed that the beam is bent in the plane containing the axis of the beam and Oy by terminal couples M. The axis Ox in the plane of the cross section is the neutral axis for elastic behaviour of the beam.

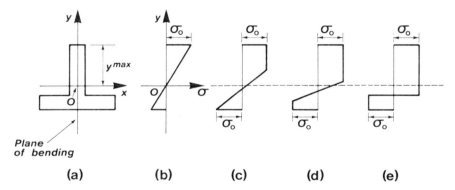

Fig. 1.7 *Stress distributions in beam with single axis of symmetry*

    (a) Cross section
    (b) At yield moment
    (c) At attainment of yield stress on lower face
    (d) Plastic zones spreading inwards from both faces
    (e) At plastic moment

In this case yield first occurs at the upper surface of the beam, as shown in Fig. 1.7(b), where y has its greatest value y^{max}. The yield moment is given by

$$M_{\text{y}} = \left(\frac{I}{y^{\text{max}}}\right)\sigma_0 = Z\sigma_0.$$

As M increases above M_{y}, a yield zone develops in the upper portion of the beam. Fig. 1.7(c) shows the corresponding stress distribution for the particular case in which the yield stress σ_0 is just attained on the lower surface of the beam. The neutral axis no longer passes through the centroid O, but assumes a position dictated by the fact that the resultant normal force on the cross section must be zero.

A further increase of bending moment causes yield to spread inwards from the lower surface of the beam, as well as spreading farther in from the upper surface, as shown in Fig. 1.7(d). Ultimately the two zones of yield meet, the distribution of stress then being as shown in Fig. 1.7(e). This is the condition of full plasticity, and the corresponding bending moment is the plastic moment.

1.3.3 Effect of upper yield stress

The theory was modified to include the effect of an upper yield stress by Robertson and Cook (1913), who assumed the stress-strain relation of Fig. 1.3(a). One consequence of this assumption is that the yield moment becomes $Z\sigma_u$, rather than $Z\sigma_0$, and the (M, κ) relation is also changed. However, as the curvature tends to infinity, the stress distributions still tend towards those shown in Fig. 1.6(c) and 1.7(e). The plastic moment, which is calculated from such fully plastic stress distributions, is therefore independent of the value of the upper yield stress.

The simple theory of elastic-plastic bending which has been outlined in this section cannot be regarded as a completely accurate description of the behaviour of steel members. The discontinuous nature of the yielding process, as indicated by the development of Lüders' lines, invalidates some of the assumptions made. For a detailed study of the problem, reference should be made to the work of Leblois and Massonet (1972).

1.4 Evaluation of plastic moment

The value of the plastic moment can be calculated directly. Fig. 1.8 shows a cross section with a single axis of symmetry Oy which lies in the plane of bending. Since the resultant axial force is zero the neutral axis in the fully plastic condition must divide the cross section into two equal areas, so that the resultant axial tensile and compressive forces are both equal to $\frac{1}{2}A\sigma_0$, where A is the total area of the cross section. If the two equal areas into which the cross section is divided have centroids G_1 and G_2 at distances $\bar{y}_1$ and $\bar{y}_2$ from the neutral axis respectively, as in Fig. 1.8, the resultant forces will act through G_1 and G_2 and the plastic moment will be given by

$$M_p = \frac{1}{2}A(\bar{y}_1 + \bar{y}_2)\sigma_0. \qquad (1.9)$$

Thus, if the plastic section modulus Z_p is defined by the relation $M_p = Z_p\sigma_0$, it follows that

$$Z_p = \frac{1}{2}A(\bar{y}_1 + \bar{y}_2). \qquad (1.10)$$

For a rectangular section of breadth B and depth D, bent about an axis parallel to the sides of breadth B, the area $A = BD$ and $\bar{y}_1 = \bar{y}_2 = D/4$, so that the plastic moment is

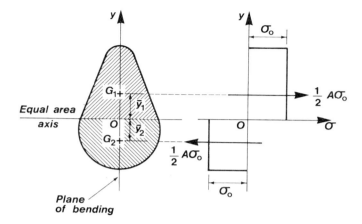

Fig. 1.8 *Fully plastic stress distribution for cross section with single axis of symmetry*

$$M_p = \tfrac{1}{4}BD^2 \sigma_0 \qquad (1.11)$$

as found previously. This result was first obtained by de Saint-Venant (1864). As already pointed out (Equation (1.4)), the yield moment is

$$M_y = \tfrac{1}{6}BD^2 \sigma_0.$$

For this cross section the ratio M_p/M_y is thus 1.5. In general, the ratio M_p/M_y is termed the shape factor, and is denoted by ν, so that

$$\nu = \frac{M_p}{M_y} = \frac{Z_p}{Z}. \qquad (1.12)$$

This ratio depends solely on the shape of the cross section.

A commercial I-section can be idealized by regarding the flanges as rectangles of breadth B and thickness T and the web as a rectangle of depth $(D - 2T)$ and thickness t, as shown in Fig. 1.9. For this idealized section it can be shown that for bending about the major axis XX, the elastic and plastic section moduli, Z and Z_p, are given by

$$Z = \frac{1}{D}[\tfrac{1}{3}BT^3 + BT(D - T)^2 + \tfrac{1}{6}t(D - 2T)^3]. \qquad (1.13)$$

$$Z_p = BT(D - T) + \tfrac{1}{4}t(D - 2T)^2. \qquad (1.14)$$

Taking as an illustration a 356×127 universal beam at 39 kg/m, with $D = 356$ mm and $B = 127$ mm, it is found from section tables that the average flange and web thicknesses are $T = 10.7$ mm and $t = 6.5$ mm, respectively. From Equations (1.13) and (1.14) it is then found that $Z = 569.3 \text{ cm}^3$ and $Z_p = 651.2$

cm^3. The values given in section tables are 570.0 cm^3 and 651.8 cm^3, differing only slightly from those calculated because of the small departures of the actual section from the ideal assumed. The shape factor derived from either pair of values is 1.14, a value typical for a rolled I-section. Formulae for the values of Z_p and ν for some of the commoner structural sections are given in Table 1.2.

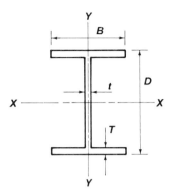

Fig. 1.9 *Idealized* I-*section*

If the axis of the applied bending moment is not parallel to, or perpendicular to, an axis of symmetry of the cross section, the plane of bending will not generally be perpendicular to the neutral (equal area) axis. A general treatment of this problem has been given by Brown (1967), and the case of a rectangular cross section was dealt with by Harrison (1963).

The plastic moment represents a definite limit on the value of the bending moment, regardless of the possible presence of residual stresses induced, for example, by previous bending into the partially plastic range. This follows from the fact that the longitudinal stress cannot exceed σ_0; on this basis the fully plastic stress distribution clearly corresponds to the greatest possible bending moment which can be developed. Moreover, it is only when this distribution is attained that the curvature can become infinite, so that a plastic hinge can form, for with any other stress distribution there must be an elastic core with a correspondingly finite rate of change of strain with distance from the neutral axis. It follows that irrespective of any residual stress distribution across a section before loading, a plastic hinge can only form when the plastic moment is attained, as pointed out by Baker and Horne (1951).

The foregoing analysis assumed that the only stresses acting were the longitudinal normal stresses due to bending. However, there will usually be shear and axial forces acting at a cross section in addition to the bending moment. These will modify the value of the plastic moment to an extent which is often negligible, but these effects may be calculated and allowed for where necessary. A

Table 1.2 *Plastic section moduli and shape factors for structural sections*

Section		Z_p	ν
Solid rectangular		$\frac{1}{4}BD^2$	1.5
Rectangular hollow section		$BT(D-T) + \frac{1}{2}T(D-2T)^2$	$B = D$ $T = 0.05\,D$ $\nu = 1.18$
Solid circular		$\frac{1}{6}D^3$	$\frac{16}{3\pi} = 1.70$
Circular hollow section		$\frac{1}{6}D^3\left[1 - \left(1 - \frac{2T}{D}\right)^3\right]$ $T \ll D; TD^2$	$T = 0.05\,D$ $\nu = 1.34$ $T \ll D$ $\nu = \frac{4}{\pi} = 1.27$
Approximation to I-section		Axis XX $BT(D-T) + \frac{1}{4}t(D-2T)^2$	About 1.14 for universal beams
		Axis YY $\frac{1}{2}TB^2 + \frac{1}{4}(D-2T)t^2$	About 1.60 for universal beams

full discussion of these and other related effects will be given in Chapter 6.

Since the value of the plastic moment is proportional to the lower yield stress σ_0, those factors which affect σ_0 will also affect the value of the plastic moment. Thus, its value will depend on such influences as the composition and heat treatment of the material, the rate of loading, and strain-ageing. These factors will also be discussed in Chapter 6.

1.5 Plastic hinge assumption for other structural materials

The plastic methods can be applied to frames of any material, provided that the members behave reasonably closely in accordance with the plastic hinge assumption. This means that whenever the bending moment reaches a critical value a plastic hinge forms and can undergo extensive rotation while the bending moment remains sensibly constant.

Reinforced concrete members often exhibit a limiting moment at which quite large hinge rotations develop, but eventually the moment falls off with further increase of rotation. The applicability of the plastic methods to reinforced concrete frames therefore depends crucially on the rotation capacity or ductility of the hinges.

A. L. L. Baker (1970) has described an ultimate load method for designing reinforced concrete frames, which differs from the simple plastic methods in that it requires only limited rotations at the plastic hinges. The rotation capacity of hinges in reinforced concrete beams has been studied extensively; a paper by Cranston and Reynolds (1970) shows that in certain circumstances a good measure of ductility is available. It has also been shown, by Cranston and Cracknell (1969), that for rectangular portal frames the simple plastic theory can provide close estimates of actual collapse loads.

References

Baker, A.L.L. (1970), *Limit State Design of Reinforced Concrete,* 2nd ed., Cement and Concrete Association, London.

Baker, J.F. (1949), 'A review of recent investigations into the behaviour of steel frames in the plastic range', *J. Inst. Civil Engrs,* **31**, 188.

Baker, J.F. and Horne, M.R. (1951), 'The effect of internal stresses on the behaviour of members in the plastic range', *Engineering,* **171**, 212.

Baker, J.F., Horne, M.R. and Heyman, J. (1956), *The Steel Skeleton,* vol. 2, Cambridge University Press.

Bauschinger, J. (1886), 'Die Veränderungen der Elastizitätsgrenze', *Mitt. mech.-tech. Lab, tech. Hochschule,* München.

Brown, E.H. (1967), 'Plastic asymmetrical bending of beams', *Int. J. Mech. Sci.,* **9**, 77.

Cranston, W.B. and Cracknell, J.A. (1969), *Tests on Reinforced Concrete Frames. 2: Portal frames with fixed feet,* Tech. Rep. TRA 420, Cement and Concrete Association, London.

Cranston, W.B. and Reynolds, G.C. (1970), *The Influence of shear on the Rotation Capacity of Reinforced Concrete Beams,* Tech. Rep. TRA 439, Cement and Concrete Association, London.

Harrison, H.B. (1963), 'The plastic behaviour of mild steel beams of rectangular section bent about both principal axes', *Struct. Engr,* **41**, 231.

Hill, R. (1950), *Plasticity,* IV, Oxford University Press, London.

Horne, M.R. (1972), 'Plastic design of unbraced sway frames', A.S.C.E.-

I.A.B.S.E. Conference on Planning and Design of Tall Buildings, Lehigh, 1972.

Leblois, C. and Massonet, C. (1972), 'Influence of the upper yield stress on the behaviour of mild steel in bending and torsion', *Int. J. Mech. Sci.,* **14**, 95.

Maier-Leibnitz, H. (1929), 'Versuche mit eingespannten und einfachen Balken von I-Form aus St. 37', *Bautechnik,* **7**, 313.

Morrison, J.L.M. (1939), 'The yield point of mild steel with particular reference to the size of specimen', *Proc. Inst. Mech. Engrs,* **142**, 193.

Robertson, A. and Cook, G. (1913), 'Transition from the elastic to the plastic state in mild steel', *Proc. R. Soc.,* A, **88**, 462.

Roderick, J.W. and Phillipps, I.H. (1949), 'The carrying capacity of simply supported mild steel beams', *Research (Eng. Struct. Suppl.) Colston Papers,* **2**, 9.

Roderick, J.W. and Heyman, J. (1951), 'Extension of the simple plastic theory to take account of the strain-hardening range', *Proc. Inst. Mech. Engrs,* **165**, 189.

Saint-Venant, B. de (1864), Article in *Resumé des Leçons,* 3rd ed. by L. Navier, Dunod, Paris.

Saint-Venant, B. de (1871), *J. Math. pures appl.* (deuxième serie), **16**, 373.

Wood, R.H. (1972), 'Rigid-jointed multi-storey steel frame design', A.S.C.E.-I.A.B.S.E. Conference on Planning and Design of Tall Buildings, Lehigh, 1972.

W.R.C.-A.S.C.E. Joint Committee, *Plastic Design in Steel − A Guide and Commentary,* 1971.

Examples

1. Show that the plastic section modulus for a solid circular section, diameter D, is $D^3/6$.

2. A rectangular hollow section has external dimensions $B = 200$ mm, $D = 400$ mm and a wall thickness $T = 12.5$ mm. Find its plastic section modulus for bending about the major axis parallel to the sides of breadth B.

3. A T-section may be regarded as composed of two rectangles, the flange being 180 mm × 15 mm and the web 165 mm × 15 mm. For bending about an axis perpendicular to the web the elastic section modulus is 124.6 cm^3. Find the corresponding value of the shape factor.

4. A beam of solid square cross section, side B, is composed of a material whose yield stresses in tension and compression are σ_0 and 1.5 σ_0, respectively. It is bent about an axis parallel to one of the sides. Find the position of the neutral axis in the fully plastic stress distribution, and the plastic moment.

5. Show that a beam of the cross section of example 3 and the material of example 4 has two different plastic moments, depending on whether the tip of the web is in tension or compression, and find their ratio.

6. A beam of solid rectangular cross section, breadth B and depth D, is bent about an axis parallel to the sides of breadth B. If the bending moment M is steadily increased to $0.88\,M_p$, find the depth of the elastic core, assuming no upper yield stress. If M is then reduced to zero, find the greatest residual stress, assuming elastic behaviour on unloading. Verify that M could then vary between $0.88\,M_p$ and $-0.453\,M_p$ without further yield taking place.

7. A uniform beam ABCD has a plastic moment M_p and is simply supported at its ends A and D:

$$AB = BC = l; \quad CD = 2l.$$

It carries a concentrated load kW at B and a second concentrated load W at C. Find the value of W which would cause plastic collapse for the three cases $k = 1$, $2, 3$.

8. A rectangular hollow section has all four sides of length B and thickness T. Find the plastic moment for bending about an axis XX passing through its centroid and parallel to two of the sides, assuming that T is very small as compared with B.

Bending moments M_x and M_y are applied about XX and an axis YY perpendicular to XX, respectively, M_x being greater than M_y. Find the relationship between M_x and M_y in the fully plastic condition.

2 Simple Cases of Plastic Collapse

2.1 Introduction

The plastic hinge hypothesis, which forms the basis of the calculation of plastic collapse loads, is summarized in Fig. 2.1, which shows the relationship between bending moment M and curvature κ for a beam of elastic flexural rigidity EI and plastic moment M_p. If the shape factor ν were unity, so that $M_y = M_p$, the beam would behave elastically until the plastic moment was attained, and then the curvature could grow indefinitely, permitting the formation of a plastic hinge. A reduction of bending moment below the plastic moment M_p would cause elastic unloading. This behaviour, shown by the full line, is the basis of the calculations given in this chapter. If M_y is less than M_p, the appropriate modification is shown schematically by the broken line in Fig. 2.1; a detailed discussion of this type of (M, κ) relation is deferred until Chapter 5.

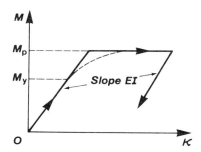

Fig. 2.1 *Ideal bending moment-curvature relation*

When a statically indeterminate frame is subjected to steadily increasing loads, the formation of the first plastic hinge does not in general cause plastic collapse. Further increases of the loads can usually be carried, and other plastic hinges form successively until finally there are enough hinges to permit a mechanism motion. Plastic collapse then occurs.

This process will be examined for a number of simple structures by means of step-by-step calculations. As will be seen in Chapter 3, it is possible to determine the plastic collapse load and the corresponding collapse mechanism for a given structure and loading by direct methods, in which no consideration is given to the sequence in which the plastic hinges form as the loads are brought up to the values which cause collapse. Indeed, the simplicity of the plastic methods is due to the fact that direct calculations of this kind can be made. Nevertheless, a thorough understanding of the process by which those plastic hinges which participate in the collapse mechanism are successively formed under steadily increasing loads is an essential preliminary to the study of the plastic methods themselves. The step-by-step technique is also used to demonstrate two important facts concerning plastic collapse loads, namely, that within wide limits the value of the plastic collapse load is unaffected by the presence of residual stresses or by the order in which the various load components are brought up to the values which cause collapse.

2.2 Simply supported beam

The first structure to be considered is a simply supported beam of uniform cross section, which has a span l and is subjected to a central concentrated load W, as shown in Fig. 2.2(a). The bending moment diagram for this beam is shown in Fig. 2.2(b), the maximum sagging bending moment at the centre of the beam being $Wl/4$. Since the beam is statically determinate, the form of this diagram is independent of the properties of the beam, and in particular of the assumed (M, κ) relation.

If W is increased steadily from zero, the beam at first behaves elastically. Eventually the central bending moment reaches the value M_p, and a plastic hinge forms beneath the load. The beam then continues to deflect at constant load as the plastic hinge rotates, and so fails by plastic collapse. The plastic collapse load W_c is determined by equating the magnitude of the central bending moment to the plastic moment, giving

$$\tfrac{1}{4} W_c l = M_p$$

$$W_c = \frac{4M_p}{l}. \qquad (2.1)$$

Since the bending moment at every cross section except the central section is less than M_p, the beam remains elastic everywhere except at the centre. The constancy of the load, and therefore of the bending moments during plastic collapse, implies constancy of the curvatures. The increases of deflection during collapse are therefore due solely to the rotation at the central plastic hinge. This effect is illustrated in Fig. 2.2(c) and (d). Curve (i) in Fig. 2.2(c) is the deflected form of the beam just as the collapse load W_c is attained, but before any rotation has occurred at the central plastic hinge. Curve (ii) is the deflected form of the beam after the central hinge has undergone rotation through an arbitrary angle 2θ. The curved shape of each half of the beam is the same in case (ii) as in case (i).

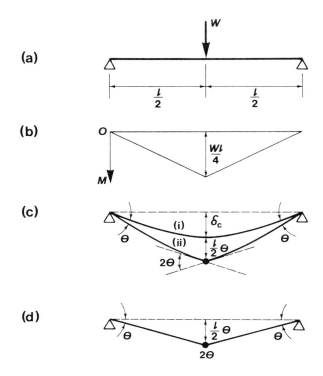

Fig. 2.2 *Simply supported beam with central concentrated load*

    (a) Loading
    (b) Bending moment diagram
    (c) (i) Elastic deflected form, $W = W_c$
       (ii) Deflected form during collapse
    (d) Changes of deflection during collapse
       Plastic hinges shown thus ●

Fig. 2.2(d) shows the changes of deflection which have occurred during plastic collapse, obtained as the difference between the deflections in case (ii) and case (i); each half of the beam is straight in this figure. These changes of deflection are thus due solely to the rotation at the plastic hinge. Fig. 2.2(d) represents the collapse mechanism for this simple case.

The elastic central deflection δ of the beam is $Wl^3/48EI$. As the collapse load is attained the central deflection δ_c at the point of collapse is therefore given by

$$\delta_c = \frac{W_c l^3}{48EI} = \frac{M_p l^2}{12EI},$$

making use of Equation (2.1). The behaviour of the beam can now be summarized on a diagram relating the load W to the central deflection δ. This load-deflection relation is shown as Ocb in Fig. 2.3. Oc is the behaviour in the elastic range, and cb represents plastic collapse under constant load, the increase of deflection from c to b being $l\theta/2$, as in the mechanism of Fig. 2.2(d).

The hinge rotation and therefore the additional deflection developed during plastic collapse is indefinite. However, if very large deflections occurred, the change in geometry of the structure would affect the conditions of equilibrium, for example by enabling the load to be supported partly by direct tension in the two halves of the beam. The simple plastic theory does not concern itself with such effects; it predicts the loads at which large deflections are imminent, as at the point c in Fig. 2.3.

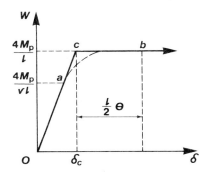

Fig. 2.3 *Load-deflection relation for simply supported beam*

The broken curve commencing at a in Fig. 2.3 shows qualitatively the effect of taking into account the difference between the yield moment M_y and the plastic moment M_p. Elastic behaviour would cease at the yield load W_y when the central bending moment was M_y, where

$$W_y = \frac{4M_y}{l} = \frac{4M_p}{vl} = \frac{W_c}{v},$$

v being the shape factor. Plastic collapse would still occur at the same value of W as before, but greater deflections would be developed before collapse. A more detailed study of this point is made in Chapter 5.

For this simple example the ratio of the collapse load W_c to the yield load W_y is equal to v, the shape factor. The ratio of W_c to W_y is always v for any statically determinate   structure, in which the greatest bending moment is proportional to the load and occurs at the same position regardless of the value of the load. Yield occurs when this greatest bending moment is equal to M_y, and collapse occurs when it is equal to M_p, for the introduction of a single hinge is always sufficient to reduce a statically determinate structure to a mechanism. It follows that the ratio of W_c to W_y is the same as the ratio of M_p to M_y, which by definition is the shape factor v.

Equation (2.1) shows how the plastic collapse load was calculated by equating the maximum bending moment to the plastic moment. This is a statical procedure but the collapse load can also be found by a kinematical procedure, as first pointed out by Horne (1949). During collapse there is no change in the elastic strain energy stored in the beam, since the bending moment distribution remains unaltered. The work done by the loads during a small motion of the collapse mechanism is therefore equal to the work absorbed in the plastic hinge, since the motion is quasi-statical. In the mechanism motion of Fig. 2.2(d) the load W_c moves through a distance $l\theta/2$ and so does work $W_c l\theta/2$. The rotation at the plastic hinge is 2θ, so that the work absorbed in the hinge is $2M_p\theta$. It follows that

$$\tfrac{1}{2}W_c l\theta = 2M_p\theta$$

$$W_c = \frac{4M_p}{l},$$

which agrees with Equation (2.1).

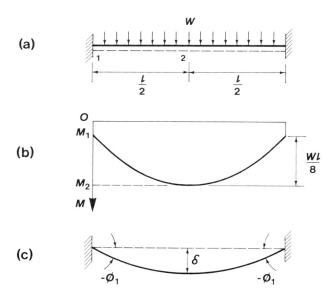

Fig. 2.4 *Fixed-ended beam with uniformly distributed load*
    (a) Loading
    (b) General form of bending moment diagram
    (c) General deflected form

2.3 Fixed-ended beam

The behaviour of a fixed-ended beam of uniform cross section and length l, carrying a uniformly distributed load W, will now be considered. The end supports are assumed to prevent rotation but permit small axial movements; if axial movement is also prevented large increases in the carrying capacity can occur, as shown by Haythornthwaite (1957).

In what follows a consistent sign convention will be used for bending moments, curvatures and hinge rotations. Positive bending moments are those which cause tensile stresses in the fibres adjacent to the broken line in Fig. 2.4(a), and positive curvatures and hinge rotations correspond to tensile strains in the same fibres.

The bending moment diagram has the parabolic form shown schematically in Fig. 2.4(b); a statical analysis gives the equilibrium equation

$$M_2 - M_1 = \frac{Wl}{8}. \qquad (2.2)$$

The beam has one statical indeterminacy or redundancy; the separate values of M_1 and M_2 cannot be found from equilibrium alone.

The state of deformation depicted in Fig. 2.4(c) forms the basis of the subsequent calculations. Here the beam has developed a slope $-\phi_1$ at each end, and the entire span is presumed to be behaving elastically. An elastic analysis (by, for example, elementary beam theory) gives the following compatibility equation:

$$M_1 = -\tfrac{1}{12}Wl - \frac{2EI\phi_1}{l}, \qquad (2.3)$$

and it can also be shown that

$$\delta = \frac{Wl^3}{384EI} - \tfrac{1}{4}l\phi_1. \qquad (2.4)$$

If W is increased steadily from zero the behaviour is at first wholly elastic, so that $\phi_1 = 0$. Equations (2.2), (2.3) and (2.4) then solve to give the elastic solution

$$M_1 = -\tfrac{1}{12}Wl$$

$$M_2 = \tfrac{1}{24}Wl$$

$$\delta = \frac{Wl^3}{384EI}.$$

Elastic behaviour ceases when $M_1 = -M_p$, so that plastic hinges form at each end of the beam. The yield load W_y is therefore given by

$$-\tfrac{1}{12}W_yl = -M_p$$

$$W_y = \frac{12M_p}{l}. \tag{2.5}$$

At this value of the load, the state of the beam is as given in the first line of Table 2.1. Fig. 2.5(a) shows the deflected form of the beam at the load W_y, and

Table 2.1 *Fixed-ended beam: proportional loading*

$\dfrac{\Delta Wl}{M_p}$	$\dfrac{Wl}{M_p}$	$\dfrac{M_1}{M_p}$	$\dfrac{M_2}{M_p}$	$\dfrac{\phi_1 EI}{M_p l}$	$\dfrac{\delta EI}{M_p l^2}$
	12	-1	0.5	0	$\frac{1}{32}$
4		0	0.5	$-\frac{1}{6}$	$\frac{5}{96}$
	16	-1	1	$-\frac{1}{6}$	$\frac{1}{12}$

Fig. 2.5(b) shows the corresponding bending moment diagram. If W increases from W_y to $W_y + \Delta W$, the plastic hinges at each end of the beam will undergo rotation while M_1 remains constant at the value $-M_p$. All changes occurring in this 'step' will be denoted by the prefix Δ. Fig. 2.5(c) shows the corresponding deflected form of the beam during this step, which is characterized by

$$M_1 = -M_p, \quad \Delta M_1 = 0, \quad \Delta\phi_1 < 0.$$

Equations (2.2), (2.3) and (2.4) become

$$\Delta M_2 = \frac{\Delta Wl}{8}. \tag{2.6}$$

$$0 = -\tfrac{1}{12}\Delta Wl - \frac{2EI\Delta\phi_1}{l}. \tag{2.7}$$

$$\Delta\delta = \frac{\Delta Wl^3}{384EI} - \tfrac{1}{4}l\Delta\phi_1. \tag{2.8}$$

Since ΔM_1 is zero, there is only one unknown bending moment increment ΔM_2, whose value is obtained immediately from the equilibrium equation (2.6). The beam is therefore statically determinate in this step. However, there is now a new geometrical unknown $\Delta\phi_1$. This is found from the compatibility equation (2.7) to be

$$\Delta\phi_1 = -\frac{\Delta Wl^2}{24EI}. \tag{2.9}$$

Substituting in Equation (2.8)

$$\Delta\delta = \frac{5\Delta Wl^3}{384EI}. \tag{2.10}$$

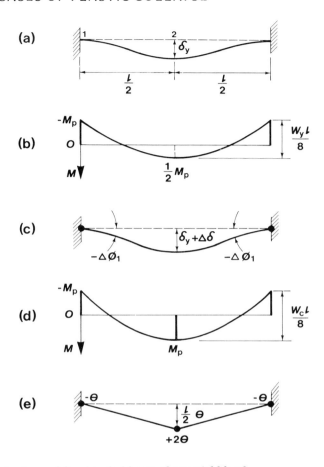

Fig. 2.5 *Behaviour of fixed-ended beam above yield load*
  (a) Deflected form, $W = W_y$
  (b) Bending moment distribution, $W = W_y$
  (c) Deflected form, $W = W_y + \Delta W$
  (d) Bending moment distribution, $W = W_c$
  (e) Collapse mechanism

Equations (2.6), (2.9) and (2.10) show that the incremental relations between ΔW, ΔM_2, $\Delta \phi_1$ and $\Delta \delta$ are those for a simply supported beam. This is because the conditions $\Delta M_1 = 0$, $\Delta \phi_1 \neq 0$ correspond to simply supported end conditions.

At the beginning of this step the value of M_2 is 0.5 M_p, as shown in Table 2.1. As ΔW increases, M_2 increases in accordance with Equation (2.6) until it reaches the value M_p. The bending moment distribution is then as shown in Fig. 2.5(d). The corresponding value of ΔW is given by

$$0.5\,M_p + \frac{\Delta Wl}{8} = M_p$$

$$\Delta W = \frac{4M_p}{l}.$$

From Equations (2.9) and (2.10) the corresponding values of $\Delta\phi_1$ and $\Delta\delta$ are

$$\Delta\phi_1 = -\frac{M_p l}{6EI}$$

$$\Delta\delta = \frac{5M_p l^2}{96EI}.$$

These incremental values are entered in the second line of Table 2.1, and the third line shows the resulting situation at the end of the step, with $W = 12\,M_p/l + 4\,M_p/l = 16\,M_p/l$.

When W has this value, a plastic hinge forms at the mid-point, and the beam then collapses, the collapse mechanism being as shown in Fig. 2.5(e). The collapse load W_c is therefore $16\,M_p/l$, a result first obtained by Kazinczy (1914).

When $W = W_c$, but before the central plastic hinge has begun to rotate, the beam is said to be at the *point of collapse*. The conditions at the point of collapse are those given in the last line of Table 2.1. The hinge rotations $-\theta$ at each end of the beam which are shown in the collapse mechanism of Fig. 2.5(e) are additional to the rotations $-M_p l/6EI$ which have already developed at the point of collapse.

The load-deflection relation is shown in Fig. 2.6, in which Oy represents the elastic behaviour up to W_y, yc represents the elastic-plastic step and cb represents plastic collapse by the mechanism of Fig. 2.5(e). The broken curve commencing at a shows schematically the type of relation which would be obtained if the yield moment M_y was less than M_p.

This load-deflection relation is typical for a beam or frame with one redundancy. When the first plastic hinge forms at the yield load (in this case a symmetrical pair of hinges), the structure is rendered statically determinate for further increases of the load, and the plastic hinge rotations which then occur cause a reduction in the slope of the load-deflection relation. Collapse does not occur until a further plastic hinge forms, thus reducing the structure to a mechanism. In general a finite increase in the load above the yield value will be required to bring the bending moment at the final plastic hinge position up to the plastic moment.

The behaviour of the fixed-ended beam is thus fundamentally different from the behaviour of the simply supported beam, for which the load-deflection relation was shown in Fig. 2.3. In that case the formation of a single plastic hinge caused collapse, and the ratio of the collapse load W_c to the yield load W_y was the shape factor ν. However, for the fixed-ended beam just considered the yield load W_y is $12M_p/vl$, while the collapse load W_c is $16M_p/l$, so that the ratio of W_c to W_y is $4v/3$. The greater margin between the yield and collapse loads for the fixed-ended beam is a consequence of the single redundancy which exists in this case.

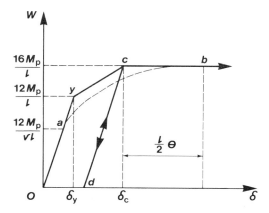

Fig. 2.6 *Load-deflection relation for fixed-ended beam*

2.3.1 *Direct calculation of collapse load*

It can be seen by inspection that there is only one possible collapse mechanism for the fixed-ended beam, this being the symmetrical mechanism of Fig. 2.5(e). This enables the plastic collapse load to be calculated directly by either a statical or a kinematical procedure.

The statical procedure consists simply of sketching the bending moment diagram at collapse, as in Fig. 2.5(d). It is seen that

$$\frac{W_c l}{8} = 2M_p$$

$$W_c = \frac{16M_p}{l}.$$

The kinematical procedure is based on the collapse mechansim of Fig. 2.5(e). Since the central deflection is $l\theta/2$, the average vertical displacement of the uniformly distributed load W_c is $l\theta/4$, so that the work done by the load during this mechanism motion is $W_c l\theta/4$. At each plastic hinge the work absorbed must be positive, and is the product of M_p and the magnitude of the hinge rotation. Equating the work done to the work absorbed,

$$\tfrac{1}{4}W_c l\theta = M_p(\theta) + M_p(2\theta) + M_p(\theta) = 4M_p\theta$$

$$W_c = \frac{16M_p}{l}.$$

2.3.2 *Behaviour on unloading*

If the load on the beam was removed after some rotation at the plastic hinges had occurred at the collapse load W_c, rotation at these hinges would cease and the behaviour during unloading would be wholly elastic, according to the assumed (M, κ) relation of Fig. 2.1. Thus if W was increased to the point of collapse c in Fig. 2.6 and then removed, the unloading line would be cd, parallel to the original elastic line Oy. The residual deflection at d can be calculated directly from the information contained in Fig. 2.6 and Table 2.1, and is $M_p l^2/24EI$.

The reason for the existence of this residual deflection is that the unloaded beam would contain residual bending moments caused by the plastic hinge rotations $-M_p l/6EI$ at each end of the beam, which stay constant during the unloading process. These residual bending moments can be calculated from Equations (2.2) and (2.3), with $W = 0$ and $\phi_1 = -M_p l/6EI$, and are found to be

$$M_1 = M_2 = \tfrac{1}{3}M_p.$$

It is readily verified that the same results are obtained by treating the unloading as a further step in which

$$\Delta W = -16M_p/l, \quad \Delta\phi_1 = 0.$$

The fact that residual moments can be induced in a structure by previous loading into the elastic-plastic range shows that the Principle of Superposition cannot be applied in such cases. For instance, if the beam were reloaded from the point d in Fig. 2.6, the beam would of necessity behave elastically along dc until the collapse load was reached at c, since the elastic behaviour during unloading is reversible. During such a reloading, the bending moments and deflections produced by a given load will be different from those arising during the first loading.

2.4 Effect of partial end-fixity

Perfect end-fixity of the kind assumed in the foregoing example (Fig. 2.4) cannot be assured in practice. To examine the effect of partial end-fixity, consider a uniform beam resting on four supports, as shown in Fig. 2.7(a). The central span is of fixed length l and carries a central concentrated load W, and the two outer spans are each of variable length kl. The non-dimensional parameter k specifies the degree of rotational constraint at the ends of the central span. With $k = 0$ the central span becomes fixed-ended, while if k is infinite this span is effectively simply supported at its ends.

The only possible collapse mechanism is as shown in Fig. 2.7(b). The collapse load W_c is found by the kinematical procedure to be given by

$$W_c \frac{l}{2}\theta = M_p(\theta) + M_p(2\theta) + M_p(\theta) = 4M_p\theta$$

$$W_c = \frac{8M_p}{l}.$$

The corresponding bending moment diagram at collapse is shown in Fig. 2.7(c). The collapse load W_c can also be found by the statical procedure, and is given by

$$\frac{W_c l}{4} = 2M_p$$

$$W_c = \frac{8M_p}{l},$$

agreeing with the kinematical calculation.

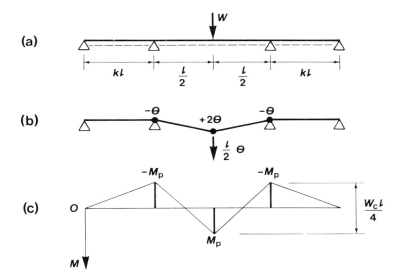

Fig. 2.7 *Continuous beam on four supports*

These analyses show that the collapse load W_c is independent of the degree of end-fixity as specified by k; its value will be $8M_p/l$, provided only that the conditions at each end of the central span are such that the plastic moment can be developed there. This illustrates one of the surprising features of the plastic theory, namely that plastic collapse loads do not depend on the actual rigidity of joints or supports.

The behaviour under a steadily increasing load can be analysed by the step-by-step process; details will not be given here. In the elastic range the greatest bending moment occurs beneath the load, and the yield load W_y and the central deflection δ_y at yield are

$$W_y = \frac{8M_p}{l}\left[\frac{3+2k}{3+4k}\right]$$

$$\delta_y = \frac{M_p l^2}{24EI}\left[\frac{3+8k}{3+4k}\right].$$

Above the yield load rotation occurs at the central plastic hinge. The bending moments at the two middle supports reach the value $-M_p$ when $W = W_c$, and the corresponding deflection δ_c at the point of collapse is

$$\delta_c = \frac{M_p l^2}{24EI}(1 + 4k).$$

Load-deflection relations derived from these results are shown in Fig. 2.8. When $k = 0$ (the fixed-ended condition), the yield load W_y coincides with the collapse load W_c, showing that in this special case all three plastic hinges form simultaneously. As k increases, both W_y and the slope of the load-deflection relation between W_y and W_c are progressively reduced, so that the deflection δ_c at the point of collapse becomes larger. However, plastic collapse always occurs at the same load $8M_p/l$ regardless of the value of k.

For values of k in excess of, say, 3, unacceptably large deflections would develop before the plastic collapse load was reached. In such cases the theoretical

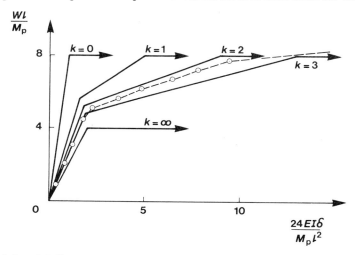

Fig. 2.8 *Load-deflection relations for beam of Fig. 2.7*

collapse load would be of little interest to the designer, since the normal purpose of calculating the collapse load is to determine the load at which large deflections are imminent. This was pointed out by Kazinczy (1934), who discussed load-deflection relations similar to those in Fig. 2.8 for a beam with partial end-fixity carrying a uniformly distributed load. When joint flexibilities are large, it is obviously necessary to supplement the calculation of the plastic collapse load by an estimate of the deflections at the point of collapse. Methods for doing this are given in Chapter 5.

In the extreme case when k is infinite, the deflection δ_c at the point of collapse also becomes infinite, and so the slope of the load-deflection relation between yield and collapse becomes zero. The load-deflection relation for this case thus appears to correspond to a collapse load of only $4M_p/l$, which would be the collapse load for a simply supported beam of span l, whereas the calculated plastic collapse load is still $8M_p/l$. However, this apparent paradox, which was pointed out by Stüssi and Kollbrunner (1935), is resolved when it is realized that the horizontal load-deflection relation which occurs in this case when $W = W_y$ is merely the limiting case in which the slope of the load-deflection relation between yield and collapse tends to zero as k tends to infinity.

Stüssi and Kollbrunner carried out tests of this kind on small beams of I-section, 4.7 cm × 3.6 cm. In these tests l was 60 cm, and values of k of 0.5, 1, 2 and 3 were used. The average of their observations from two tests of this kind with $k = 2$ are shown in Fig. 2.8, and it will be seen that the comparison between the observations and the theoretical relation is good.

Further tests of a similar nature were carried out by Maier-Leibnitz (1936). A type of test which is similar in principle is obtained by applying a central vertical load to a rectangular portal frame (as in Fig. 2.9(a) with $H = 0$), in which case the horizontal member functions as a partially fixed-ended beam. Tests of this kind have been described by Girkmann (1932), Baker and Roderick (1938), and also by Hendry (1950), who showed that increasing the height of the frame while leaving the span constant did not affect the collapse load but increased the deflections prior to collapse. Similar tests, but with symmetrical two-point loading, have been described by Rusek, Knudsen, Johnston and Beedle (1954). The effect of partial end-fixity on the design of beams subjected to uniformly distributed loads, whose ends are encased in reinforced concrete or masonry, was the subject of a theoretical and experimental investigation by Kazinczy (1934). The behaviour of a full-scale portal frame whose feet were supported by short piled footings was investigated experimentally by Baker and Eickhoff (1955), who showed that for this frame the collapse load was not affected by the partial fixity of the feet to any appreciable extent.

2.5 Rectangular portal frame

The last structure to be considered is the rectangular portal frame whose dimensions and loading are shown in Fig. 2.9(a). All the members of this frame are uniform with flexural rigidity EI and plastic moment M_p. The joints at sections 2 and 4 are rigid, and the columns are rigidly built-in at their bases 1 and 5. The sign convention for bending moment, curvature and hinge rotation is again that positive values correspond to tensile stresses or strains in the fibres adjacent to the broken line.

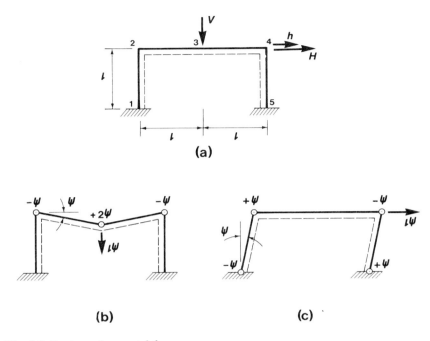

(a)

(b)   **(c)**

Fig. 2.9 *Rectangular portal frame*
     (a) Dimensions and loading
     (b) Virtual beam mechanism
     (c) Virtual sway mechanism

Within each of the four segments of the frame which are straight and free from external load, namely 12, 23, 34 and 45, the shear force must be constant. The bending moment must therefore vary linearly along each of these segments. The values of the bending moments M_1, M_2, M_3, M_4 and M_5 at the five numbered cross sections therefore specify the bending moment distribution throughout the frame. Moreover, since the bending moment cannot exceed M_p in magnitude at any cross section, it follows that plastic hinges can only occur at the ends of these segments. Thus, the only possible locations of plastic hinges are the five numbered sections. (This excludes the special case in which the shear force is zero in a segment, so that the bending moment is constant along the segment.)

The frame has three redundancies, for if a cut were made at any section, and the shear force, axial force and bending moment were all specified at this section, it would become statically determinate. It follows that there must be two equations of equilibrium connecting the five bending moments. For the determination of the bending moments which occur when the frame is wholly elastic there must therefore be three equations of compatibility. For situations in which some plastic hinges have developed, it is still possible to develop three appropriate equations of compatibility, and the two equations of equilibrium will still apply.

The equilibrium and compatibility equations will now be derived using the Principle of Virtual Work. This procedure has certain advantages, and will be used again in later chapters.

2.5.1 *Principle of Virtual Work*

The Principle of Virtual Work for frames involves *force systems* (of loads and bending moments) which satisfy the requirements of equilibrium and *displacement systems* (of deflections, curvatures and hinge rotations) which satisfy the requirements of compatibility. For the type of problem under consideration it takes the form

$$\sum P\delta = \int M\kappa \, ds + \sum M\phi. \tag{2.11}$$

In this equation M is any distribution of bending moment satisfying the requirements of equilibrium with concentrated external loads P. κ denotes any distribution of curvatures which is compatible with hinge rotations ϕ and deflections δ. The summation on the left-hand side covers all points of application of external loads. On the right-hand side the integral covers all members of the frame, distance s being measured along each member, and the summation covers all sections where there may be a hinge rotation.

The equation is valid provided that the force system (P, M) satisfies the requirements of equilibrium and the displacement system (δ, κ, ϕ) satisfies the requirements of compatibility. In addition, the sign convention for δ must be consistent with that for P, so that the direction of a positive force P must be the same as the direction of the positive corresponding displacement δ. Similarly, the sign conventions for both κ and ϕ must be consistent with that for M.

Equation (2.11) can be used in two ways. In the first of these the displacement system $(\delta^*, \kappa^*, \phi^*)$ is virtual (throughout this Section, asterisks are used to denote virtual systems, either of displacements or of forces). This means that the displacements, curvatures and hinge rotations can be chosen arbitrarily, subject only to the requirements of compatibility, and need not be ascribable to any possible form of loading. This form of the principle is often referred to as the Principle of Virtual Displacements, and its use generates equations of equilibrium.

The other possibility is to use a virtual force system (P^*, M^*). Here the forces and moments are chosen arbitrarily, subject only to the requirements of equilibrium. This is the Principle of Virtual Forces, and its application results in equations of compatibility.

Both forms of the principle will now be used to develop the equations of equilibrium and of compatibility which are needed for the step-by-step analysis of the frame of Fig. 2.9(a).

2.5.2 *Equilibrium equations by virtual displacements*

The two equilibrium equations will be derived using the virtual displacement systems depicted in Fig. 2.9(b) and (c). These represent a beam mechanism and a sway mechanism, respectively. The hinges are not plastic hinges, but are introduced to permit the small displacements which are shown to take place while the members between the hinges remain straight. Because the curvature is everywhere zero, Equation (2.11) reduces to

$$\sum P\delta^* = \sum M\phi^*.$$

The equilibrium system consists of the loads H and V shown in Fig. 2.9(a), together with bending moments at the five numbered sections which are in equilibrium with these loads. Using this system in conjunction with the virtual displacement systems of Fig. 2.9(b) and (c) in turn gives

$$Vl\psi = M_2(-\psi) + M_3(+2\psi) + M_4(-\psi)$$
$$Hl\psi = M_1(-\psi) + M_2(+\psi) + M_4(-\psi) + M_5(+\psi).$$

Cancelling ψ throughout gives the two equilibrium equations

$$Vl = -M_2 + 2M_3 - M_4. \tag{2.12}$$

$$Hl = -M_1 + M_2 - M_4 + M_5. \tag{2.13}$$

These are equations of equilibrium which must be obeyed regardless of whether the frame is behaving elastically or has become partly plastic. Equation (2.11) is independent of material properties, and these properties did not enter into the derivation of Equations (2.12) and (2.13).

2.5.3 *Compatibility equations by virtual forces*

The three compatibility equations are found by using virtual force systems in which all the external P^* are zero. The bending moments are then referred to as residual moments and denoted by the symbol m^*. Equation (2.11) becomes

$$0 = \int m^*\kappa\,ds + \sum m^*\phi. \tag{2.14}$$

The actual curvature κ at any section is related to the actual bending moment M at that section by the elastic relation $\kappa = M/EI$, so that Equation (2.14) becomes

$$0 = \int \frac{m^*M}{EI} ds + \sum m^*\phi. \tag{2.15}$$

Within each of the linear segments 12, 23, 34 and 45, both m^* and M vary linearly with distance s, along the member. In these circumstances the integral in Equation (2.15) is readily evaluated for a typical uniform straight segment AB of length L, and is

$$\int_B^A \frac{m^*M}{EI} ds = \frac{L}{6EI}[m_A^*(2M_A + M_B) + m_B^*(2M_B + M_A)]. \tag{2.16}$$

Equation (2.15), taken in conjunction with Equation (2.16), can be used to generate three compatibility equations, provided that three linearly independent residual moment systems can be identified. To achieve this, it is noted that any set of residual moments must obey the two equations of equilibrium (2.12) and (2.13), with the loads V and H each zero, giving

$$-m_2 + 2m_3 - m_4 = 0. \tag{2.17}$$

$$-m_1 + m_2 - m_4 + m_5 = 0. \tag{2.18}$$

Following Heyman (1961), the three systems used are as shown in the first three lines of Table 2.2.

Table 2.2 *Virtual force and actual displacement systems*

Section		1	2	3	4	5
Virtual force systems						
m^*	(i)	1	1	0.5	0	0
	(ii)	0	0	0.5	1	1
	(iii)	0	1	1	1	0
M^*		$-l$	0	0	0	0
Actual displacement system						
$EI\kappa = M$		M_1	M_2	M_3	M_4	M_5
ϕ		ϕ_1	ϕ_2	ϕ_3	ϕ_4	ϕ_5

It is readily verified that each of the three residual moment systems (i), (ii) and (iii) satisfies Equations (2.17) and (2.18), and that they are linearly independent.

The actual displacement system is shown in the last two lines of Table 2.2. When used in conjunction with the virtual residual moment system (i), Equations (2.15) and (2.16) give

$$\frac{l}{6EI}[(3M_1 + 3M_2) + (2.5M_2 + 2M_3) + (M_3 + 0.5M_4)] + \phi_1 + \phi_2 + 0.5\phi_3 = 0$$

and this reduces to

$$3M_1 + 5.5M_2 + 3M_3 + 0.5M_4 + 6EI(\phi_1 + \phi_2 + 0.5\phi_3)/l = 0.$$

$$(2.19)$$

Similarly, the use of the residual moment systems (ii) and (iii) gives

$$0.5M_2 + 3M_3 + 5.5M_4 + 3M_5 + 6EI(0.5\phi_3 + \phi_4 + \phi_5)/l = 0.$$

$$(2.20)$$

$$M_1 + 5M_2 + 6M_3 + 5M_4 + M_5 + 6EI(\phi_2 + \phi_3 + \phi_4)/l = 0.$$

$$(2.21)$$

Equations (2.19)–(2.21) are the three equations of compatibility. These equations, and the two equations of equilibrium (2.12) and (2.13), may be written more compactly and in incremental form as follows:

$$
\begin{bmatrix}
0 & -1 & 2 & -1 & 0 \\
-1 & 1 & 0 & -1 & 1 \\
\hdashline
3 & 5.5 & 3 & 0.5 & 0 \\
0 & 0.5 & 3 & 5.5 & 3 \\
1 & 5 & 6 & 5 & 1
\end{bmatrix}
\begin{bmatrix}
\Delta M_1 \\ \Delta M_2 \\ \Delta M_3 \\ \Delta M_4 \\ \Delta M_5
\end{bmatrix}
$$

$$
+ \frac{6EI}{l}
\begin{bmatrix}
0 & 0 & 0 & 0 & 0 \\
0 & 0 & 0 & 0 & 0 \\
\hdashline
1 & 1 & 0.5 & 0 & 0 \\
0 & 0 & 0.5 & 1 & 1 \\
0 & 1 & 1 & 1 & 0
\end{bmatrix}
\begin{bmatrix}
\Delta\phi_1 \\ \Delta\phi_2 \\ \Delta\phi_3 \\ \Delta\phi_4 \\ \Delta\phi_5
\end{bmatrix}
=
\begin{bmatrix}
\Delta Vl \\ \Delta Hl \\ 0 \\ 0 \\ 0
\end{bmatrix}
\qquad (2.22)
$$

where the prefix Δ is used to denote changes in bending moments, hinge rotations and loads. The two equilibrium equations are those whose coefficients appear above the broken line.

These five equations may be used to trace the behaviour of the frame when subjected to *proportional loading*,

$$V = H = W,$$

up to collapse, starting from a condition in which the frame is unloaded and free from stress. Because the ideal (M, κ) relation of Fig. 2.1 is assumed,

$$\text{either} \quad |M| < M_p, \quad \Delta M \neq 0, \quad \Delta\phi = 0,$$

$$\text{or} \quad |M| = M_p, \quad \Delta M = 0, \quad \Delta\phi \neq 0,$$

so that the Equations (2.22) only involve five unknowns in any step.

2.5.4 *Step-by-step analysis of proportional loading*

As W increases, the frame behaves elastically in the first instance, so that the changes of hinge rotation $\Delta\phi_1$, $\Delta\phi_2$, $\Delta\phi_3$, $\Delta\phi_4$ and $\Delta\phi_5$ are all zero. The equations then have the solution

$$\Delta M_1 = -0.2125\Delta Wl$$
$$\Delta M_2 = -0.0125\Delta Wl$$
$$\Delta M_3 = 0.3\Delta Wl$$
$$\Delta M_4 = -0.3875\Delta Wl$$
$$\Delta M_5 = 0.4125\Delta Wl.$$

The greatest bending moment is $0.4125\Delta Wl$ at section 5. This reaches the value M_p at the yield load W_y, which is given by

$$0.4125W_yl = M_p$$
$$W_y = 2.424M_p/l.$$

The distribution of bending moment is then as shown in the first line of Table 2.3.

When W increases above the yield load to, say, $W_y + \Delta W$, the plastic hinge which has formed at section 5 undergoes rotation while M_5 remains constant at the value M_p. Thus, during this step

$$M_5 = M_p, \quad \Delta M_5 = 0, \quad \Delta\phi_5 > 0$$
$$\Delta\phi_1 = \Delta\phi_2 = \Delta\phi_3 = \Delta\phi_4 = 0.$$

Equations (2.22) then have the solution

$$\Delta M_1 = -0.468\Delta Wl$$
$$\Delta M_2 = 0.108\Delta Wl$$
$$\Delta M_3 = 0.342\Delta Wl$$
$$\Delta M_4 = -0.424\Delta Wl$$
$$\Delta\phi_5 = 0.209\Delta Wl^2/EI.$$

During this step the number of redundancies has dropped from 3 to 2, because one bending moment increment is now known, namely $\Delta M_5 = 0$. There are therefore only four statical unknowns, ΔM_1, ΔM_2, ΔM_3 and ΔM_4. The two equations of equilibrium therefore need to be supplemented by only two of the equations of compatibility to determine these four bending moment increments. However, there is now a geometrical unknown, $\Delta\phi_5$, and this is found from the third equation of compatibility.

Table 2.3 Rectangular frame: proportional loading

$\dfrac{\Delta Wl}{M_p}$	$\dfrac{Wl}{M_p}$	$\dfrac{M_1}{M_p}$	$\dfrac{M_2}{M_p}$	$\dfrac{M_3}{M_p}$	$\dfrac{M_4}{M_p}$	$\dfrac{M_5}{M_p}$	$\dfrac{\phi_3 EI}{M_p l}$	$\dfrac{\phi_4 EI}{M_p l}$	$\dfrac{\phi_5 EI}{M_p l}$	$\dfrac{hEI}{M_p l^2}$
	2.424	−0.515	−0.030	0.727	−0.939	1				0.177
0.143		−0.067	0.015	0.049	−0.061	0			0.030	0.020
	2.567	−0.582	−0.015	0.776	−1	1			0.030	0.197
0.390		−0.331	0.058	0.224	0	0		−0.217	0.101	0.100
	2.957	−0.913	0.043	1	−1	1		−0.217	0.131	0.297
0.043		−0.087	−0.043	0	0	0	0.167	−0.116	0.036	0.036
	3	−1	0	1	−1	1	0.167	−0.333	0.167	0.333

A plastic hinge of negative sign forms next at section 4, where the bending moment at the start of this step was $-0.939M_p$. The value of ΔW for the step is thus given by

$$-0.939M_p - 0.424\Delta Wl = -M_p$$

$$\Delta W = 0.143M_p/l.$$

With this value of ΔW the increments which occur in this step are shown in the second line of Table 2.3, and the resulting situation is given in the third line.

During the next step rotations occur at the plastic hinges at both sections 4 and 5, so that

$$M_4 = -M_p, \quad \Delta M_4 = 0, \quad \Delta\phi_4 < 0$$

$$M_5 = M_p, \quad \Delta M_5 = 0, \quad \Delta\phi_5 > 0$$

$$\Delta\phi_1 = \Delta\phi_2 = \Delta\phi_3 = 0.$$

The corresponding solution of Equations (2.22) is

$$\Delta M_1 = -0.85\Delta Wl$$

$$\Delta M_2 = 0.15\Delta Wl$$

$$\Delta M_3 = 0.575\Delta Wl$$

$$\Delta\phi_4 = -0.558\Delta Wl^2/EI$$

$$\Delta\phi_5 = 0.258\Delta Wl^2/EI.$$

There are only three statical unknowns, ΔM_1, ΔM_2 and ΔM_3 in this step, so that the number of redundancies has dropped to one. Only one equation of compatibility needs to be used in conjunction with the two equations of equilibrium to determine these three unknowns, but the remaining two equations of compatibility are required to find the values of the two geometrical unknowns $\Delta\phi_4$ and $\Delta\phi_5$.

The next plastic hinge forms at section 3, where the bending moment at the start of this step was $0.776M_p$. The value of ΔW for the step is therefore given by

$$0.776M_p + 0.575\Delta Wl = M_p$$

$$\Delta W = 0.390M_p/l.$$

The corresponding increments during this step are given in the fourth line of Table 2.3, and the fifth line shows the situation at the end of the step.

The ensuing step is characterized by

$$M_3 = M_p, \quad \Delta M_3 = 0, \quad \Delta\phi_3 > 0$$

$$M_4 = -M_p, \quad \Delta M_4 = 0, \quad \Delta\phi_4 < 0$$

$$M_5 = M_p, \quad \Delta M_5 = 0, \quad \Delta \phi_5 > 0$$

$$\Delta \phi_1 = \Delta \phi_2 = 0.$$

The structure is now statically determinate; the two equilibrium equations suffice to determine ΔM_1 and ΔM_2, and the three compatibility equations furnish values of $\Delta \phi_3$, $\Delta \phi_4$ and $\Delta \phi_5$. The solution is

$$\Delta M_1 = -2\Delta Wl$$

$$\Delta M_2 = -\Delta Wl$$

$$\Delta \phi_3 = 3.833 \Delta Wl^2/EI$$

$$\Delta \phi_4 = -2.667 \Delta Wl^2/EI$$

$$\Delta \phi_5 = 0.833 \Delta Wl^2/EI.$$

At the end of this step a plastic hinge forms at section 1, where the bending moment at the start of the step was $-0.913 M_p$. The value of ΔW for this step is given by

$$-0.913 M_p - 2\Delta Wl = -M_p$$

$$\Delta W = 0.043 M_p/l.$$

The changes which occur during this step and the final situation are given in the sixth and seventh lines of Table 2.3.

The final load is $3M_p/l$, and at this load there are four plastic hinges. This reduces the frame to the mechanism which is illustrated in Fig. 2.10. The collapse load is thus $W_c = 3M_p/l$. Since $W_y = 2.424 M_p/l$, the ratio of W_c to W_y in this case is $3/2.424$, or 1.24. If the effect of the shape factor ν is allowed for, this ratio becomes 1.24ν.

2.5.5 Deflections by unit load method

The analysis will be completed by the calculation of a particular deflection at the end of each step. This is achieved by using an appropriate virtual force system in conjunction with the actual system of displacements. If the deflection sought is δ, all that is necessary is to make the corresponding P^* unity, with all other external loads zero. This is the well-known unit load method; reference to Equation (2.11) shows that

$$\delta = \int \frac{M^*M}{EI} \, ds + \sum M^* \phi, \tag{2.23}$$

where M^* represents any distribution of bending moments satisfying the requirements of equilibrium with the virtual unit load.

For example, let h be the horizontal deflection at section 4 (Fig. 2.9(a)) corresponding to H. Examination of the equilibrium equations (2.12) and (2.13)

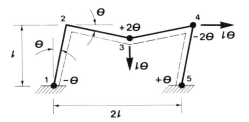

Fig. 2.10 *Collapse mechanism for frame of Fig. 2.9(a); V = H = W*

shows that the simple system of bending moments $M_1^* = -l, M_2^* = M_3^* = M_4^* = M_5^* = 0$, satisfies the requirements of equilibrium with loads $H = 1$ and $V = 0$. This system, which is entered in the fourth line of Table 2.2, can therefore be used in Equation (2.23) to calculate h. The integral in Equation (2.23) can be evaluated using Equation (2.16), with M^* replacing m^*, and the result which is obtained is

$$h = -\frac{l^2}{6EI}(2M_1 + M_2) - l\phi_1. \qquad (2.24)$$

Values of h are given in Table 2.3, and Fig. 2.11(a) shows the relationship between W and h for proportional loading. Each point at which a fresh plastic hinge forms is marked with the number of the relevant cross section, in accordance with Fig. 2.9(a).

2.6 Invariance of collapse loads

If a structure is subjected to more than one load, the loads will only rarely increase in proportion to one another. Fortunately, wide variations can occur in the manner in which the various loads are brought up to their collapse values without affecting the collapse load. Thus for the frame just considered, collapse occurs under proportional loading when $H = V = 3M_p/l$. To take an extreme case of loading which is not proportional, suppose that a load $V = 3M_p/l$ is first applied, and then held constant while H is increased steadily from zero. A step-by-step analysis shows that the load $V = 3M_p/l$ is borne by wholly elastic action, and that H must then be increased to the value $2.133M_p/l$ before the first plastic hinge forms. This hinge forms at section 4, although under proportional loading the first hinge formed at section 5. The load-deflection relation is shown in Fig. 2.11(b). When a fourth plastic hinge is formed at section 1 the value of H is $3M_p/l$; collapse then occurs by the same mechanism as before. Thus the collapse load condition for this case, $H = V = 3M_p/l$, is the same as for proportional loading. The only difference between the two cases is in respect of the load-deflection relation prior to collapse.

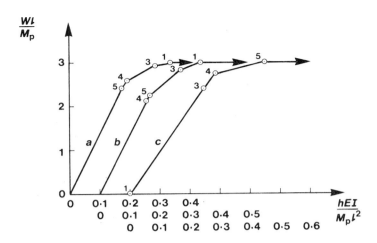

Fig. 2.11 *Load-deflection relations for frame of Fig. 2.9(a)*

(a) Proportional loading: $H = W$, $V = W$
(b) $V = 3M_\text{p}/l$, followed by $H = W$
(c) Feet spread apart, followed by proportional loading

The plastic collapse load is not affected by residual stresses, whether these be due to welding, imperfect fit of members, plastic hinge rotations which have occurred during previous loading or the movements of supports. To illustrate this point, suppose that in the frame of Fig. 2.9(a) the bases had spread apart, while the columns remained fixed in direction, the spread being just sufficient to cause the bending moments at the bases 1 and 5 to reach the value $-M_\text{p}$ in the unloaded condition. An elastic analysis shows that the beam is then subjected to a uniform sagging bending moment $M_\text{p}/3$. If the loads H and V are then increased proportionately, rotation of a plastic hinge at section 1 begins immediately. However, the bending moment at section 5 undergoes a positive change, so that there is no immediate plastic hinge rotation at this section. The load-deflection relation is shown in Fig. 2.11(c). It is again found that collapse occurs when $H = V = 3M_\text{p}/l$, although the load-deflection relation differs from the other two cases, and the sequence of formation of hinges is also quite different.

The fact that residual stresses have no effect on the value of the plastic collapse load for a given structure was pointed out by Kazinczy (1938). Direct experimental confirmation of this point was provided by the work of Maier-Leibnitz (1928) and of Horne (1952) on continuous beams, in which the effect of initial lowering of supports on the collapse load was shown to be negligible. Moreover, any welded frame will contain residual stresses due to the welding process unless a stress-relieving treatment is used, so that indirect confirmation has been supplied by the lack of any noticeable effect in the many tests which have been carried out on such structures.

The invariance of the plastic collapse load stems from the fact that collapse can occur only when a sufficient number of plastic hinges have formed to transform the structure into a mechanism. The rectangular frame just considered has three redundancies. If at any stage of the loading there are three plastic hinges, at each of which the bending moment is known, the frame becomes statically determinate, so that the remaining bending moments can be calculated in terms of the loads from the equations of equilibrium. If a fourth plastic hinge then forms, the frame is reduced to a mechanism, and the knowledge of the value of a fourth bending moment enables the loads to be calculated, provided of course that the value of each load is specified as a multiple of a single load W. Thus once the mechanism of collapse is known, the collapse load W_c can be calculated by considering only the equations of equilibrium. These equations are independent of residual stresses, the order of application of loads, the imperfect rigidity of joints or the sinking of supports, and so the collapse load is unaffected by such factors.

To illustrate this point, consider the collapse mechanism of Fig. 2.10. The bending moments at the plastic hinge positions are

$$M_1 = -M_p, \quad M_3 = M_p, \quad M_4 = -M_p, \quad M_5 = M_p.$$

Substituting these values in the two equations of equilibrium (2.12) and (2.13), with $H = V = W_c$, it is found that

$$-M_2 + 3M_p = W_c l$$
$$M_2 + 3M_p = W_c l$$

so that $W_c = 3M_p/l$ and $M_2 = 0$. This value of W_c agrees with the value found by the step-by-step calculations.

The collapse load can also be calculated directly by a kinematical procedure. Equating the work done to the work absorbed at the plastic hinges for the mechanism of Fig. 2.10 gives

$$W_c l\theta + W_c l\theta = M_p(\theta) + M_p(2\theta) + M_p(2\theta) + M_p(\theta)$$
$$2W_c l\theta = 6M_p\theta$$
$$W_c = 3\frac{M_p}{l}.$$

As will be seen from these calculations, the determination of the collapse load once the actual collapse mechanism is known is remarkably swift. However, the collapse mechanism cannot usually be foreseen unless the structure is extremely simple. As already remarked, in the case of the fixed-ended beam of Fig. 2.4 there was only one possible collapse mechanism, which must therefore be the actual collapse mechanism. However, in the example just considered, there are three possible collapse mechanisms. One of these is the mechanism of Fig. 2.10, which was shown by the step-by-step analysis to be the actual collapse mechanism. The other two possibilities are the mechanisms of Fig. 2.9(b) and (c), with plastic hinges in place of the virtual hinges shown. If the step-by-step calculations had not been performed it would not be known *a priori* which of these three mechanisms was the actual collapse mechanism. In more complicated frames there is a much wider choice of possible collapse mechanisms, and so it is necessary to have some guiding principles to enable the actual collapse mechanism to be identified. The available principles will be stated and discussed in Chapter 3.

2.7 Plastic design

It is now widely recognized that the rational approach to the design of a structure is based on considering the various *limit states* at which the structure becomes unfit for use. Some limit states are associated with *serviceability,* such as the development of deflections large enough to damage internal finishes in buildings. Others are concerned with catastrophic collapse, due for example to fatigue failure, brittle fracture or buckling. One such *ultimate limit state* for steel frames is plastic collapse, and there are many practical cases in which this limit state governs the design. In such cases a plastic design procedure is appropriate.

Once the shape of a frame and the lengths of its members are decided, the *characteristic loads* can be determined. Different types of loading will be known with different degrees of accuracy. For instance, dead loading can be determined more precisely than imposed loading. The pressure distributions due to a given incident wind speed and direction are difficult to assess, and the wind speed itself has to be selected on the basis of statistical data (CP3 (1972) chapter V).

A margin of safety is provided by multiplying the characteristic loads by *load factors.* Different load factors may be used for different types of loads and their possible combinations. The task of the designer is then to select the cross sections of the members so that plastic collapse would just occur under the factored loads. With an appropriate choice of load factors, the probability of plastic collapse actually occurring is made sufficiently remote to be acceptable. Load factors are usually specified in relevant Codes of Practice.

When plastic design methods are appropriate, they have a clear advantage over elastic methods. This is apparent from a consideration of the results obtained for the frame shown in Fig. 2.9(a). Under proportional loading, $H = V = W$, it was found that $W_c/W_y = 1.24\nu$. Taking $\nu = 1.14$, a typical value for an I-section, $W_c/W_y = 1.41$.

In this particular case, the plastic collapse load is thus 41 per cent greater than the load at which yield first occurs, when elastic analysis ceases to be valid. A design based on elastic analysis could not make use of this additional reserve of strength beyond the elastic limit, and would therefore be less economical than a plastic design. The margin of strength which exists above W_y depends on the particular structure and loading, but the example given is fairly typical. Plastic design is therefore to be preferred on the score of economy, quite apart from the obvious rationality of basing the design on a direct consideration of the relevant ultimate limit state.

A further advantage of plastic design which has emerged is that the plastic collapse load is independent of such factors as imperfect rigidity of joints, the movement of supports and the presence of residual stresses, all of which have a profound effect on the elastic stress distribution. Finally, plastic collapse analysis is much simpler than elastic analysis, as will become apparent in Chapters 3 and 4.

References

Baker, J.F. and Eickhoff, K.G. (1955), 'The behaviour of saw-tooth portal frames', prelim vol., Conf. Correlation between Calculated and Observed Stresses and Displacements in Structures, Inst. Civil Engrs, 107.

Baker, J.F. and Roderick, J.W. (1938), 'An experimental investigation of the strength of seven portal frames', *Trans. Inst. Weld.,* **1**, 206.

CP3, Chapter V (1972): Loading Part 2: Wind Loads, B.S.I.

Girkmann, K. (1932), 'Über die Auswirkung der "Selbsthilfe" des Baustahls in Rahmenartigen Stabwerken', *Stahlbau,* **5**, 121.

Haythornthwaite, R.M. (1957), 'Beams with full end fixity', *Engineering,* **183**, 110.

Hendry, A.W. (1950), 'An investigation of the strength of certain welded portal frames in relation to the plastic method of design', *Struct. Engr,* **28**, 311.

Heyman, J. (1961), 'On the estimation of deflexions in elastic-plastic framed structures', *Proc. Inst. Civil Engrs,* **19**, 39.

Horne, M.R. (1949), Contribution to 'The design of steel frames' by Baker, J.F., *Struct. Engr,* **27**, 421.

Horne, M.R. (1952), 'Experimental investigations into the behaviour of continuous and fixed-ended beams', prelim. publ., 4th Congr. Inst. Assoc. Bridge Struct. Eng., Cambridge, 1952, 147.

Kazinczy, G.v. (1914) 'Kisérletek befalazott tartókkal', *Betonszemle,* **2**, 68.

Kazinczy, G.v. (1934), 'Die Bemessung unvollkommen eingespannter Stahl I-Deckenträger unter Berücksichtigung der plastischen Formänderungen', *Proc. Int. Assoc. Bridge Struct. Eng.,* **2**, 249.

Kazinczy, G.v. (1938), 'Versuche mit innerlich statisch unbestimmten Fachwerken', *Bauingenieur,* **19**, 236.

Maier-Leibnitz, H. (1928), 'Beitrag zur Frage der tatsächlichen Tragfähigkeit einfacher und durchlaufender Balkenträger aus Baustahl St. 37 und aus Holz', *Bautechnik,* **6**, 11.

Maier-Leibnitz, H. (1936), 'Versuche zur weiteren Klärung der Frage der tatsächlichen Tragfähigkeit durchlaufender Träger aus Baustahl', *Stahlbau,* **9**, 153.

Rusek, J.M., Knudsen, K.E., Johnston, E.R. and Beedle, L.S. (1954), 'Welded portal frames tested to collapse', *Weld. J., Easton, Pa.,* **33**, 469-s.

Stüssi, F. and Kollbrunner. C.F. (1935), 'Beitrag zum traglastverfahren', *Bautechnik*, **13**, 264,

Examples

1. A uniform beam AB of length l and plastic moment M_p is simply supported at its ends A and B. Calculate the collapse load by both the statical and kinematical methods for the following loadings:

(a) a uniformly distributed load W
(b) a concentrated load W at a distance $l/3$ from A.

2. For the beam of example 1, loading case (b), use the kinematical method to find the collapse load if the end conditions were altered as follows:

(a) A: simply supported   B: fixed-ended
(b) A: fixed-ended   B: simply supported
(c) A and B both fixed-ended

3. A uniform beam of length l and plastic moment M_p is fixed at both ends. It carries a uniformly distributed load W and also a central concentrated load W. Find the value of W which would cause plastic collapse.

4. A uniform beam of length l and plastic moment M_p is fixed at both ends. It carries three concentrated loads, each of magnitude W, at distances $l/4$, $l/2$ and $3l/4$ from one end. What value of W would cause plastic collapse?

5. A rectangular portal frame ABCD has vertical columns AB and DC each of height h and the horizontal beam BC is of span l. The feet A and D are rigidly built-in, and the joints B and C are rigid. All members of the frame are uniform, the plastic moment being M_p. A concentrated vertical load W is applied to the beam at a distance μl from the joint B. Find the value of W at which plastic collapse would occur.

3 Basic Theorems and Simple Examples

3.1 Introduction

As shown in Chapter 2, the plastic collapse load can be calculated very simply once the mechanism is known. For a few simple structures there is only one possible collapse mechanism, but in most cases this is not so. There is therefore a need for theorems which enable the actual collapse mechanism to be selected from among the various possibilities. The purpose of this chapter is to state these theorems and give some examples of their application. A description of some general methods for the determination of plastic collapse loads is deferred until Chapter 4.

The basic assumption is made that whenever the plastic moment M_p is attained in a member, a plastic hinge forms which can undergo rotation of any magnitude, provided that the bending moment stays constant. For the present it is assumed that the value of M_p is a definite constant for a given member, and does not depend on the axial and shear forces which the member may be called upon to sustain. In fact, the value of the plastic moment is affected by axial and shear forces, and also by such factors as the local stress concentrations which occur beneath the points of application of concentrated loads. However, these effects are often negligibly small, and their discussion is deferred until Chapter 6.

It is also assumed that the deflections of a frame under consideration are small enough for the equations of statical equilibrium to be sensibly the same as those for the undistorted frame, an assumption which also underlies conventional methods of elastic analysis.

3.2 Statement of theorems

For each of the simple examples considered in Chapter 2, it was evident that when the collapse load was reached a sufficient number of plastic hinges had formed to transform the structure into a mechanism. The deflections could then increase under constant load due to rotations occurring at these hinges, while the bending moments at the hinges remained constant at their fully plastic values. From a consideration of the requirements of statical equilibrium it followed that the bending moment distribution throughout the structure stayed unchanged during collapse. It was also evident that during plastic collapse the work done by the external loads was equal to the work absorbed in the plastic hinges. These results were obviously true for the simple cases considered, but it is desirable that they should be established for the general case. Formal proofs were supplied by Greenberg (1949) using the terminology of truss-type structures; an adaptation of his argument to the case of framed structures is given in Appendix A.

3.2.1 *Static theorem*

In general, there will exist many distributions of bending moment throughout a redundant frame which satisfy all the conditions of statical equilibrium with a prescribed set of external loads. Greenberg and Prager (1952) termed distributions of this kind *statically admissible*. In addition, a distribution of bending moment in which the plastic moment is not exceeded anywhere in the frame is described as *safe*. A necessary condition for a frame to be capable of carrying a given set of loads is evidently that there must be at least one safe distribution of bending moment throughout the frame which is statically admissible with these loads. The static theorem states that this condition is also sufficient to ensure that the frame can carry the loads.

For a formal statement of the theorem, suppose that a frame is subjected to loads λP_1, λP_2, ..., λP_n, each load being applied at a given point in a specified direction. P_1, P_2, ..., P_n are presumed fixed, and may be thought of as the characteristic loads. λ is then the load factor. The loads are specified completely by the value of λ, and can be referred to collectively as the set of loads λ. The load factor which would cause plastic collapse is denoted by λ_c, this value of λ being termed the *collapse load factor*. The static theorem can now be stated as follows:

Static theorem. If there exists any distribution of bending moment throughout a frame which is both safe and statically admissible with a set of loads λ, the value of λ must be less than or equal to the collapse load factor λ_c.

A corollary of this theorem is that if for a given set of loads λ it can be shown that no distribution of bending moment exists which is both safe and statically admissible, this value of λ must be greater than the collapse load factor λ_c. It follows that a frame can actually carry the highest loads which could conceivably be carried without collapse, since λ_c is the highest load factor at which statical equilibrium can be maintained without the plastic moment being exceeded somewhere in the frame.

The static theorem was first suggested by Kist (1917) as an intuitive axiom. A proof was supplied by Gvozdev (1936), and later by Greenberg and Prager (1952) and also by Horne (1950), and is given in Appendix A.

A further corollary concerns the effect of strengthening a frame by increasing the plastic moment of one or more of the members. This cannot result in a decrease of the collapse load factor. Thus if a frame will collapse at a load factor λ_c, there must be at least one distribution of bending moment which is safe and statically admissible with the set of loads λ_c. This same distribution of bending moment must remain safe and statically admissible with these loads if the plastic moment is increased at one or more cross sections, for the requirements of statical equilibrium remain unchanged, and if the plastic moment was not exceeded anywhere in the original frame it will certainly not be exceeded in the strengthened frame. This result was stated by Feinberg (1948) as an axiom, no proof being offered.

3.2.2 *Kinematic theorem*

If the actual collapse mechanism is known for a given frame and loading, the collapse load factor can be found by equating the work done by the loads during a small motion of the collapse mechanism to the work absorbed in the plastic hinges. When the actual collapse mechanism is not known, a work equation of this kind can be written down for any assumed mechanism. A value of λ will then be obtained which *corresponds* to the assumed mechanism. The kinematic theorem is concerned with such corresponding values of λ, and can be stated as follows:

Kinematic theorem. For a given frame subjected to a set of loads λ, the value of λ which corresponds to any assumed mechanism must be either greater than or equal to the collapse load factor λ_c.

A corollary is that if the values of λ corresponding to all the possible collapse mechanisms are determined, the actual collapse load factor λ_c will be the smallest of these values.

A formal proof of this theorem is given in Appendix A. It was established by Gvozdev (1936), and also by Greenberg and Prager (1952), who used a physical argument based on Feinberg's axiom which itself has been shown to be a direct consequence of the static theorem. Consider a particular frame A subjected to a set of loads λ, the collapse value of λ being λ_c. Let any mechanism be assumed and the work equation written down, giving a corresponding value of λ. Now imagine another frame B which is derived from frame A by increasing the plastic moments indefinitely at all cross sections except those where plastic hinges occur in the assumed mechanism, where they remain unaltered. The actual collapse mechanism for the strengthened frame B must be the mechanism originally assumed for frame A, since no other mechanism is possible. The actual collapse load factor for frame B, obtained from the work equation, must therefore be λ. By Feinberg's axiom the collapse load factor λ_c for frame A cannot be greater than the collapse load factor λ for the strengthened frame B, and this establishes the kinematic theorem.

3.2.3 Uniqueness theorem

The static and kinematic theorems can be combined to form a uniqueness theorem. Thus, it is known from the static theorem that for any value of λ above λ_c, there is no distribution of bending moment which is both safe and statically admissible. Moreover, it is known from the kinematic theorem that there is no mechanism for which the corresponding load factor is less than λ_c. Combining these results, the following theorem can be stated:

Uniqueness theorem. For a given frame and set of loads λ, if there is at least one safe and statically admissible bending moment distribution, in which the plastic moment occurs at enough cross sections to produce a mechanism, the corresponding load factor will be the collapse load factor λ_c.

This theorem was proved by Horne (1950). A convenient summary of this and the two earlier theorems is:

$$\left. \begin{array}{ll} \text{Statical conditions} & \lambda \leqslant \lambda_c \\ \text{Kinematical conditions} & \lambda \geqslant \lambda_c \end{array} \right\} \ \lambda = \lambda_c$$

where the statical and kinematical conditions are those specified in the two corresponding thorems.

3.3 Illustrative example

The significance of the theorems will now be discusssed, using as an example the frame whose dimensions and loading are shown in Fig. 3.1(a). Each member of this frame is uniform and has a plastic moment 25 kN m; it is required to find the collapse load factor λ_c. The sign convention for bending moments and hinge rotations will again be that positive values correspond to tensile stresses or strains in the fibres adjacent to the broken line.

The frame itself, although not the loads, is identical with the frame discussed in Section 2.5 (see Fig. 2.9). As pointed out in that Section, plastic hinges can only occur at the five numbered cross sections shown in Fig. 3.1(a). Moreover, there are only three possible collapse mechanisms; these are shown in Fig. 3.1(b), (c) and (d).

The small motion of the sway mechanism depicted in Fig. 3.1(b) is completely defined by the clockwise rotation of the left-hand column, and the hinge rotations and the horizontal displacement of the beam are as shown. Fig. 3.1(c), which shows the kinematics of the beam mechanism, is self-explanatory. The third mechanism, shown in Fig. 3.1(d), results if the motions of the sway and beam mechanisms are added, and is referred to as the combined mechanism.

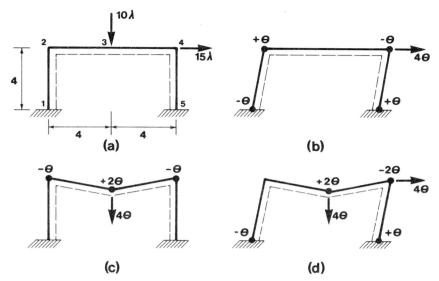

Fig. 3.1 *Rectangular portal frame*

    (a) Dimensions and loading. Units: m, kN
    (b) Sway mechanism
    (c) Beam mechanism
    (d) Combined mechanism

3.3.1 *Equilibrium equations by virtual displacements*

In discussing this problem, use will be made of the equations of equilibrium, which are obtained using the Principle of Virtual Displacements, as explained in Section 2.5.2. The equilibrium system consists of the applied loads and any set of bending moments M_1, M_2, M_3, M_4 and M_5 which satisfies the requirements of equilibrium with these loads.

The frame has three redundancies, and so there must be two equations of equilibrium relating these five bending moments to the applied loads. These two equations are derived using as virtual displacement systems the sway and beam mechanisms of Fig. 3.1(b) and (c), for this purpose regarding the hinges not as plastic hinges, but as virtual hinges which permit the small virtual movements defined.

Equating the virtual work done by the loads to the virtual work absorbed at the hinges in each of the two mechanisms,

$$60\lambda\theta = M_1(-\theta) + M_2(+\theta) + M_4(-\theta) + M_5(+\theta). \tag{3.1}$$

$$60\lambda = -M_1 + M_2 - M_4 + M_5. \tag{3.2}$$

$$40\lambda\theta = M_2(-\theta) + M_3(+2\theta) + M_4(-\theta). \tag{3.3}$$

$$40\lambda = -M_2 + 2M_3 - M_4. \tag{3.4}$$

3.3.2 *Sway mechanism*

Consider the sway mechanism, Fig. 3.1(b). A kinematical analysis will first be carried out, in which the mechanism is treated as though it were the actual collapse mechanism. During the small motion of this mechanism which is shown, the horizontal load 15λ moves through a distance 4θ and so does work $60\lambda\theta$. (For convenience, units are not stated in the analysis, as they are self-evident.) To the first order of small quantities the beam does not move vertically, and so the vertical load does no work. At each plastic hinge the work absorbed must be positive, regardless of the sign of the hinge rotation. There are four plastic hinges, at each of which the rotation is of magnitude θ, so that the work absorbed is 25θ. Equating the work done to the work absorbed,

$$60\lambda\theta = 25\theta + 25\theta + 25\theta + 25\theta = 100\theta. \tag{3.5}$$

$$\lambda = 1.667. \tag{3.6}$$

If this were the actual collapse mechanism, this would be the value of λ_c. In fact, the calculation gives the value of λ corresponding to this mechanism. According to the kinematic theorem, this value of λ is either greater than or equal to λ_c,

$$\lambda_c \leqslant 1.667, \tag{3.7}$$

so that the analysis establishes this upper bound on the value of λ_c.

Now consider a statical analysis. If this were the actual collapse mechanism, the bending moments at the four plastic hinges would each be of magnitude 25, and their signs would be the same as the signs of the plastic hinges, so that

$$M_1 = -25, \quad M_2 = +25, \quad M_4 = -25, \quad M_5 = +25. \tag{3.8}$$

These are the four bending moments appearing in Equation (3.2). Substituting in Equation (3.1), from which Equation (3.2) was derived, it is found that

$$60\lambda\theta = -25(-\theta) + 25(+\theta) - 25(-\theta) + 25(+\theta)$$
$$= 25\theta + 25\theta + 25\theta + 25\theta = 100\theta, \tag{3.9}$$

and this is identical with the kinematical Equation (3.5).

This identity between the statical and kinematical analyses is both general and of fundamental importance. In this particular case it stems from the fact that the equation of equilibrium, Equation (3.1), is derived from the sway mechanism of Fig. 3.1(b), treated as a virtual mechanism, and the work equation (3.5) is derived from the same mechanism treated as an assumed plastic collapse mechanism. The plastic moments involved, Equations (3.8), must all be of the same sign as the corresponding plastic hinge rotation, as is seen in Equation (3.9). It follows that all the products on the right-hand side of this equation must be positive. This is the equivalent of the physical statement used in the kinematical analysis, namely that the work absorbed at a plastic hinge is always positive, regardless of the sense of its rotation.

A further important deduction can be made from Equation (3.2),

$$60\lambda = -M_1 + M_2 - M_4 + M_5. \tag{3.2}$$

Each of the four bending moments appearing on the right-hand side of this equation must lie between the limits ± 25. The highest conceivable value which λ could have, viewed from the standpoint of this equation alone, is therefore obtained by setting

$$M_1 = -25, \quad M_2 = +25, \quad M_4 = -25, \quad M_5 = +25.$$

These are the values which correspond to the sway mechanism, as specified in Equations (3.8). It can be concluded that

$$60\lambda \leqslant 100; \quad \lambda \leqslant 1.667.$$

This argument thus leads to the same conclusion as the kinematic theorem, namely that the value of λ corresponding to the assumed mechanism could not be exceeded, and is therefore an upper bound on λ_c. It also shows that the sway mechanism corresponds to the *breakdown* of the equilibrium equation (3.2), for this equation cannot be satisfied if λ exceeds 1.667 without one or more bending moments exceeding the plastic moment 25 in magnitude.

These arguments can be given generality. A kinematical analysis of an assumed mechanism gives an upper bound on λ_c, and there is always a corresponding equilibrium equation whose breakdown, in the sense defined, gives the same upper bound.

A full statical analysis of the sway mechanism must involve both equilibrium equations (3.2) and (3.4). The four plastic moments occurring in the sway mechanism are given in Equations (3.8). When these values are substituted in Equation (3.2), it is found that $\lambda = 1.667$, as previously shown. Equation (3.4) then gives

$$2M_3 = 40\lambda + M_2 + M_4$$
$$= 40 \times 1.667 + 25 - 25 = 66.7$$
$$M_3 = 33.3.$$

Since M_3 cannot exceed the plastic moment 25 in magnitude, it follows that the sway mechanism cannot be the actual collapse mechanism.

This statical analysis shows that the set of bending moments

$$M_1 = -25$$
$$M_2 = +25$$
$$M_3 = +33.3$$

$$M_4 = -25$$

$$M_5 = +25$$

is statically admissible with the applied loads when $\lambda = 1.667$. Since the equations of equilibrium are linear in the bending moments and loads, it follows that the set of bending moments obtained by multiplying the above set by any positive factor k will be statically admissible with the loads corresponding to a load factor $k\lambda$. If k is chosen as 0.75, the corresponding set of bending moments becomes safe, since M_3 is reduced to $+25$. Thus, the set of bending moments

$$M_1 = -18.75$$

$$M_2 = +18.75$$

$$M_3 = +25$$

$$M_4 = -18.75$$

$$M_5 = +18.75$$

is statically admissible with the loads defined by $\lambda = 0.75 \times 1.667 = 1.25$, and is also safe. It follows from the static theorem that

$$\lambda_c \geqslant 1.25.$$

This procedure for deriving a lower bound is due to Greenberg and Prager (1952). Combining this lower bound with the upper bound already obtained,

$$1.25 \leqslant \lambda_c \leqslant 1.667.$$

3.3.3 Beam mechanism

The beam mechanism is shown in Fig. 3.1(c). Treating this mechanism as the actual collapse mechanism, a kinematical analysis gives the corresponding value of λ. Since there is no horizontal movement of the columns to the first order of small quantities, the horizontal load does no work. The vertical load 10λ moves through a distance 4θ and so does work $40\lambda\theta$. There are three plastic hinges at which the magnitudes of the rotations are θ, 2θ and θ. The work equation is therefore:

$$40\lambda\theta = 25\theta + 25(2\theta) + 25\theta = 100\theta$$

$$\lambda = 2.5. \tag{3.10}$$

By the kinematic theorem,

$$\lambda_c \leqslant 2.5.$$

This upper bound is higher than the value found from the sway mechanism. It can therefore be concluded that the beam mechanism cannot be the actual collapse mechanism. Nevertheless, a statical analysis will now be given to provide a further illustration of the use of the static theorem.

From Fig. 3.1(c) the bending moments at the three plastic hinges are

$$M_2 = -25, \quad M_3 = +25, \quad M_4 = -25. \tag{3.11}$$

These are the three bending moments appearing in Equation (3.4), the equilibrium equation which was derived from the beam mechanism. Substituting in Equation (3.3), from which Equation (3.4) was derived,

$$40\lambda\theta = -25(-\theta) + 25(+2\theta) - 25(-\theta)$$

$$= 25\theta + 25(2\theta) + 25\theta = 100\theta.$$

This equation is again identical with the kinematical equation (3.10). It also follows that the beam mechanism corresponds to the breakdown of Equation (3.4).

To complete the statical analysis, the value of λ just found is used in the other equation of equilibrium, Equation (3.2), giving

$$M_5 - M_1 = 60\lambda - M_2 + M_4$$

$$= 60 \times 2.5 + 25 - 25 = 150. \tag{3.12}$$

M_5 and M_1 are not determined uniquely. This is because the number of unknowns in the original problem is four, namely the three redundancies and the value of λ. However, the three plastic hinges in the beam mechanism provide only three items of statical information. Since both M_5 and M_1 cannot exceed the plastic moment 25 in magnitude, Equation (3.12) cannot be satisfied by any safe values. The best lower bound is arrived at by taking

$$M_1 = -75, \quad M_5 = +75.$$

These two bending moments, together with the three plastic hinge values given in Equations (3.11), constitute a bending moment distribution which is statically admissible with $\lambda = 2.5$. If each of these moments, and the value of λ, is multiplied by $1/3$, the distribution

$$M_1 = -25$$
$$M_2 = -8.33$$
$$M_3 = +8.33$$
$$M_4 = -8.33$$
$$M_5 = +25$$

is obtained, which is statically admissible with $\lambda = 2.5/3 = 0.833$. Since this distribution is also safe, this value of λ is a lower bound on λ_c. Combining this with the upper bound previously obtained,

$$0.833 \leqslant \lambda_c \leqslant 2.5.$$

These bounds are not as close as those resulting from the analysis of the sway mechanism.

3.3.4 Combined mechanism

This mechanism is shown in Fig. 3.1(d). For the kinematical analysis it is seen that both horizontal and vertical loads 15λ and 10λ move through the same distance 4θ, so that the total work done is $100\lambda\theta$. Equating this to the work absorbed at the plastic hinges,

$$100\lambda\theta = 25\theta + 25(2\theta) + 25(2\theta) + 25\theta = 150\theta$$

$$\lambda = 1.5. \tag{3.13}$$

This is an upper bound on λ_c, by the kinematic theorem, so that

$$\lambda_c \leqslant 1.5.$$

This is the lowest of the upper bounds obtained from the three possible collapse mechanisms. It can be concluded that this must be the actual value of λ_c, and that the combined mechanism is the actual collapse mechanism.

For the statical analysis, it is seen from Fig. 3.1(d) that the bending moments at the four plastic hinges are

$$M_1 = -25, \quad M_3 = +25, \quad M_4 = -25, \quad M_5 = +25. \tag{3.14}$$

Substituting these values of bending moments in the two equations of equilibrium (3.2) and (3.4), it is found that

$$\lambda = 1.5, \quad M_2 = 15.$$

M_2 is less than the plastic moment in magnitude, so that the bending moment distribution found by this analysis is both safe and statically admissible with $\lambda = 1.5$. Since there are enough plastic hinges to constitute a mechanism, the requirements of both the static and kinematic theorems have been met. By the uniqueness theorem, it follows that

$$\lambda_c = 1.5.$$

It is instructive to note that if the equilibrium equations (3.1) and (3.3) are added, the following equation of equilibrium is obtained

$$100\lambda\theta = M_1(-\theta) + M_3(+2\theta) + M_4(-2\theta) + M_5(+\theta). \tag{3.15}$$

This equation only involves the bending moments at the four plastic hinges occurring in the combined mechanism, and its breakdown is expressed as follows

$$100\lambda\theta = -25(-\theta) + 25(+2\theta) - 25(-2\theta) + 25(+\theta) = 150\theta,$$

corresponding to the substitution of the plastic hinge moments involved in this mechanism, Equations (3.14). The addition of Equations (3.1) and (3.3) to form Equation (3.15) is the statical counterpart of the kinematical addition of the beam and sway mechanisms to form the combined mechanism. It is readily confirmed that Equation (3.15) can also be derived by applying the Principle of Virtual Displacements to the combined mechanism treated as a virtual mechanism.

In designing a frame, the problem is to determine the values of the plastic moments of the members so that a specified load factor against collapse is provided. For the frame of Fig. 3.1(a), it has now been established that if each member has a plastic moment 25 kN m, λ_c is 1.5. If, for example, a load factor of 1.6 was specified, the analysis has shown that this would require the plastic moment to be increased to

$$\left(\frac{1.6}{1.5}\right)25 = 26.7 \text{ kN m}.$$

In the discussion of this example, two possible methods for the determination of collapse loads have been revealed. The first method is to assume a mechanism of collapse and to carry out a complete statical analysis for this mechanism, thus obtaining the complete bending moment distribution. If this distribution is such that the plastic moment is not exceeded anywhere in the frame, the corresponding load factor must be the actual collapse load factor, by the uniqueness theorem. If not, other mechanisms of collapse are analysed similarly until the correct collapse mechanism is found. This method, which may be termed the *trial-and-error method*, was first proposed by Baker (1949), and an example of its use is given in Chapter 4. The second method is to examine all the possible collapse mechanisms, writing down a work equation for each mechanism and thus deriving the corresponding value of the load factor. The collapse load factor will then be the smallest value thus obtained, by the kinematic theorem. For simple frames it is comparatively easy to carry out the necessary computations, but in fairly complicated frames there would clearly be a great number of possible mechanisms and the procedure would become very tedious. A method for circumventing the necessity for analysing all possible mechanisms is described in Chapter 4.

Much experimental work has been carried out on rectangular portal frames subjected to horizontal and vertical loading, and it has been found that the collapse loads predicted by the plastic theory are usually in excellent agreement with the observed loads at which large deflections are imminent. The most comprehensive series of tests were those reported by Baker and Heyman (1950) on miniature frames, and confirmation of these results was provided by some full-scale tests described by Baker and Roderick (1952). Further full-scale tests have been reported by Schilling, Schutz and Beedle (1956).

3.4 Distributed loads

When a member in a frame is subjected to a uniformly distributed load, the bending moment distribution is parabolic, and a maximum bending moment may then occur at any position. If the actual collapse mechanism involves a plastic hinge at the position of maximum bending moment, the location of this hinge will need to be determined. The calculation of the collapse load is thus more lengthy in such cases, although good approximations can be found by employing the technique of upper and lower bounds.

3.4.1 *Maximum bending moment in a member*

As a preliminary step, the position and magnitude of the maximum bending moment in a member will be stated. Fig. 3.2 shows a member of length L carrying a total load W which is uniformly distributed. It is supposed that the bending moments M_C, M_L and M_R are known, C being the centre of the member and L and R the left- and right-hand ends, respectively. The maximum bending moment $M^{\max}$ occurs at a position specified alternatively by x_0, y_0 or z_0.

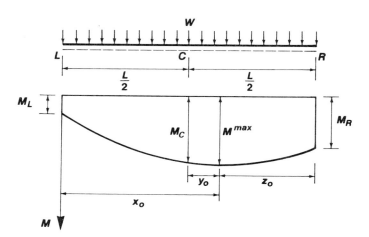

Fig. 3.2 *Bending moment diagram for member carrying a uniformly distributed load*

The following results can be established by elementary statics:

$$x_0 = (4M_C - 3M_L - M_R)/W.$$
$$y_0 = (M_R - M_L)/W. \qquad (3.16)$$
$$z_0 = (4M_C - M_L - 3M_R)/W.$$

$$M^{\max} = M_L + Wx_0^2/2L$$
$$= M_C + Wy_0^2/2L \qquad (3.17)$$
$$= M_R + Wz_0^2/2L.$$

The position and magnitude of $M^{\max}$ will be found most accurately by calculating the smallest of the three distances x_0, y_0 or z_0.

There may be no position of maximum bending moment within the span, so that the bending moment increases or decreases continuously from one end of the member to the other. This would be the case if, for example, it was found that z_0 was negative.

3.4.2 *Illustrative example*

Consider the rectangular frame whose dimensions and loading are shown in Fig. 3.3(a), the total load on the beam being 48λkN, uniformly distributed. The members are all uniform, with plastic moment 40 kN m.

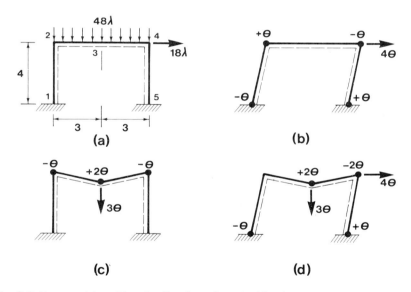

Fig. 3.3 *Frame with uniformly distributed vertical load*
  (a) Dimensions and loading. Units: m, kN
  (b) Sway mechanism
  (c) Beam mechanism
  (d) Combined mechanism

The initial assumption, later to be corrected, is that if a plastic hinge occurs anywhere in the beam it will be at the centre, section 3. With this assumption the three possible collapse mechanisms are shown in Fig. 3.3(b), (c) and (d); these are similar to those shown in Fig. 3.1. The two equations of equilibrium can be derived treating the mechanisms of Fig. 3.3(b) and (c) as virtual mechanisms. For the sway mechanism the horizontal load 18λ moves through a distance 4θ and so does work $72\lambda\theta$, and the vertical load does no work. The virtual work equation is thus

$$72\lambda\theta = M_1(-\theta) + M_2(+\theta) + M_4(-\theta) + M_5(+\theta)$$
$$72\lambda = -M_1 + M_2 - M_4 + M_5. \tag{3.18}$$

In the beam mechanism the horizontal load does no work. The centre of the beam moves vertically through a distance 3θ. The *average* vertical movement of the uniformly distributed load 48λ is thus 1.5θ, so that the work done by this load is $72\lambda\theta$. The virtual work equation is

$$72\lambda\theta = M_2(-\theta) + M_3(+2\theta) + M_4(-\theta)$$
$$72\lambda = -M_2 + 2M_3 - M_4. \tag{3.19}$$

For each of the three possible collapse mechanisms the bending moments at the plastic hinges can be written down by inspection from Fig. 3.3(b), (c) and (d). Substitution in the relevant equations of equilibrium then gives the corresponding values of λ. The values thus obtained are

$$\text{Sway} \qquad \lambda = 2.222$$
$$\text{Beam} \qquad \lambda = 2.222$$
$$\text{Combined} \quad \lambda = 1.667.$$

From the kinematic theorem, it follows that 1.667 is an upper bound on the value of λ_c, and that the combined mechanism is the actual collapse mechanism, subject to the correct location of the plastic hinge which was assumed to occur at mid-span. This mechanism is now investigated in more detail.

The bending moments at the assumed plastic hinges are:

$$M_1 = -40, \quad M_3 = +40, \quad M_4 = -40, \quad M_5 = +40. \tag{3.20}$$

Substituting in Equations (3.18) and (3.19), it is found that

$$\lambda = 1.667, \quad M_2 = 0.$$

Equations (3.16) and (3.17) may now be used to determine the greatest bending moment in the beam. Reference to Fig. 3.2 shows that for this case

$$M_L = M_2 = 0$$
$$M_C = M_3 = +40 \qquad W = 48 \times 1.667 = 80$$
$$M_R = M_4 = -40 \qquad L = 6.$$

Using these values it is found that

$$y_0 = -0.5$$

$$M^{\max} = 40\left(\frac{25}{24}\right) = 41.67.$$

The complete statical solution for the mechanism of Fig. 3.3(d) thus gives rise to a bending moment distribution which is statically admissible with $\lambda = 1.667$ but is not safe. If the bending moments and λ are each multiplied by the factor $24/25 = 0.96$, a safe and statically admissible bending moment distribution results, as follows:

λ	1.667	1.6
M_1	-40	-38.4
M_2	0	0
$M^{\max}$	$+41.67$	$+40$
M_4	-40	-38.4
M_5	$+40$	$+38.4$

This establishes a lower bound of 1.6 upon the value of λ_c. Combining this with the upper bound already found,

$$1.6 \leqslant \lambda_c \leqslant 1.667. \qquad (3.21)$$

This result defines the value of λ_c to within ± 2 per cent. If the exact value is required, the precise location of the plastic hinge in the beam must be found. This may be done by a kinematical analysis of the mechanism shown in Fig. 3.4, in which the hinge in the beam is positioned at a distance y to the right of the centre of the beam. Any value of y between the limits ± 3 defines a mechanism, for which the corresponding value of λ is an upper bound on λ_c. By the kinematic theorem, the lowest of these upper bounds must be the actual collapse load factor.

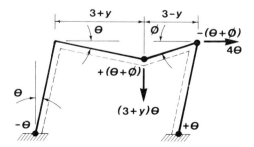

Fig. 3.4 *Combined mechanism. Units:* m

The motion of the mechanism is defined by the clockwise rotation θ of each column. This is also the rotation of the left-hand portion of the beam of length $(3 + y)$, so that the vertical deflection at the plastic hinge within the beam is $(3 + y)\theta$. The right-hand portion of the beam of length $(3 - y)$ rotates counter-clockwise through an angle ϕ, where

$$(3 - y)\phi = (3 + y)\theta. \tag{3.22}$$

The average vertical deflection of the uniformly distributed load 48λ on the beam is $0.5(3 + y)\theta$, so that the work done by this load is $24\lambda(3 + y)\theta$. The horizontal load 18λ moves through a distance 4θ and so does work 72θ. Equating the work done to the work absorbed,

$$24\lambda(3 + y)\theta + 72\lambda\theta = 2 \times 40\theta + 2 \times 40(\theta + \phi).$$

Eliminating ϕ with the aid of Equation (3.22), it is found that

$$\lambda = \frac{10}{3}\left[\frac{9 - y}{(6 + y)(3 - y)}\right]. \tag{3.23}$$

The value of y which minimizes λ is -0.487 m, and the corresponding value of λ, which is the actual collapse load factor λ_c, is 1.645.

The correct value of y differs only slightly from the value $y_0 = -0.5$ m, which defined the position of maximum bending moment in the beam when a plastic hinge was assumed to occur at the mid-point. For all practical purposes it would be sufficiently accurate to assume that the hinge in the beam is in fact located at this position, 0.5 m to the left of the mid-point, and to determine λ_c by a kinematical analysis of the mechanism thus defined. This procedure gives a value of λ_c of 1.645, which is the same as the exact value to four significant figures.

3.5 Partial and overcomplete collapse

Consider a frame with r redundancies, for which the collapse mechanism has only one degree of freedom with $(r + 1)$ plastic hinges. At collapse the values of $(r + 1)$ bending moments at the plastic hinge positions are known, and there will be one equation of equilibrium corresponding to the collapse mechanism from which the collapse load factor may be determined. There are thus r equations of equilibrium remaining from which the r redundancies can be found, so that the entire frame is statically determinate at collapse. This situation is described as *complete collapse*. The problems considered in Sections 3.3 and 3.4 were of this type, the frames having three redundancies and four plastic hinges in the collapse mechanisms.

When the collapse is not complete in this sense, it may be either *partial* or *overcomplete*. Partial collapse is said to occur if the plastic hinges which are formed in the collapse mechanism do not render the entire frame statically determinate at collapse. The term overcomplete collapse is used when there are two or more mechanisms for which the corresponding value of the load factor is the same, this value being the actual collapse load factor λ_c. Examples of both these situations will now be given.

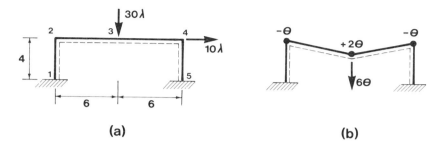

(a)   (b)

Fig. 3.5 *Example of partial collapse*
    (a) Dimensions and loading. Units: m, kN
    (b) Collapse mechanism

3.5.1 *Example of partial collapse*

A simple case of partial collapse is shown in Fig. 3.5(a) and (b). The members of this frame are of uniform cross section throughout, with plastic moment 70 kN m. The equations of equilibrium may be derived as before by the virtual displacements method, and are

$$180\lambda = -M_2 + 2M_3 - M_4 \tag{3.24}$$

$$40\lambda = -M_1 + M_2 - M_4 + M_5. \tag{3.25}$$

The beam mechanism of Fig. 3.5(b) is found to have the lowest corresponding load factor and is therefore the actual collapse mechanism. The bending moments at the three plastic hinges are

$$M_2 = -70, \quad M_3 = +70, \quad M_4 = -70. \tag{3.26}$$

Substituting these values in the two equations of equilibrium it is found that

$$\lambda_c = 1.556$$

$$M_5 - M_1 = 62.2. \tag{3.27}$$

The values of M_1 and M_5 are thus not uniquely determined for this mechanism. This is because for this frame $r = 3$, and there are only three plastic hinges in the collapse mechanism.

It is easy to select pairs of values of M_1 and M_5 which satisfy Equation (3.27) and in addition do not exceed the plastic moment 70 kN m in magnitude. Two such pairs are $M_1 = 0$, $M_5 = 62.2$; $M_1 = -31.1$, $M_5 = 31.1$. Any such pair of moments, together with those specified in Equations (3.26), would be safe and statically admissible with $\lambda_c = 1.556$. From the uniqueness theorem this confirms that the collapse mechanism is the beam mechanism of Fig. 3.5(b), and that the collapse load factor is 1.556.

In general, a part of the frame becomes statically determinate in a case of partial collapse, owing to the formation of a mechanism. It is always possible to calculate the load factor λ corresponding to an assumed partial mechanism by writing down the work equation for this mechanism, or equivalently by using the corresponding equation of equilibrium. It is not then necessary to determine the actual bending moment distribution in the rest of the structure at collapse. So long as *any* bending moment distribution can be found which is both safe and statically admissible with the set of loads λ, it is known that collapse must occur by this partial mechanism, and that $\lambda = \lambda_c$.

In the present example it can be shown that if the idealized bending moment-curvature relation of Fig. 2.1 is assumed, the values of M_1 and M_5 at collapse under proportional loading would be 3.9 and 66.1, respectively, but a knowledge of these moments is not required to decide that the actual collapse mechanism is the beam mechanism.

3.5.2 *Example of overcomplete collapse*

Consider the frame shown in Fig. 3.6(a), in which each column has a plastic moment 40 kN m, while the beam has a plastic moment 60 kN m, as indicated in the figure. This is the first example that has been encountered of a frame whose members do not all have the same plastic moment. The only additional consideration which this introduces is that if a plastic hinge forms at either of the joints 2 or 4, it will occur in the corresponding column, with a plastic moment 40 kN m, rather than in the stronger beam.

A kinematical approach will be adopted. For the sway mechanism illustrated in Fig. 3.6(b), the work equation is

$$20\lambda(5\theta) = 40\theta + 40\theta + 40\theta + 40\theta$$

$$100\lambda\theta = 160\theta$$

$$\lambda = 1.6. \tag{3.28}$$

For the combined mechanism of Fig. 3.6(c), the work equation is

$$30\lambda(2.5\phi) + 20\lambda(5\phi) = 40\phi + 60(2\phi) + 40(2\phi) + 40(\phi)$$

$$175\lambda\phi = 280\phi$$

$$\lambda = 1.6. \tag{3.29}$$

The same value of λ thus corresponds to both the sway and combined mechanisms, and it is readily verified that a higher value of λ, namely 2.67, corresponds

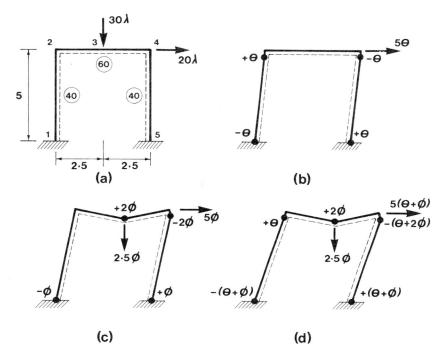

Fig. 3.6 *Example of overcomplete collapse*

    (a) Dimensions and loading. Units: m, kN

      Plastic moments shown thus: $\boxed{M_p}$

    (b) Sway mechanism
    (c) Combined mechanism
    (d) Mechanism with two degrees of freedom

to the beam mechanism. From the kinematic theorem it follows that the value of the collapse load factor λ_c is 1.6, but it appears that the collapse mechanism could be either the sway or the combined mechanism.

    These results imply that the correct description of the collapse mechanism in this case is as indicated in Fig. 3.6(d), in which the displacements and hinge rotations of the two mechanisms of Fig. 3.6(b) and (c) have been added, forming a mechanism with two degrees of freedom specified by the angles θ and ϕ. Here the only restrictions on θ and ϕ are that these angles should be positive, so that the hinge rotations at sections 2 and 3 are in the correct sense. The relative magnitudes of θ and ϕ need not be specified. The corresponding work equation, as derived directly from a consideration of the kinematics of this mechanism, is

$$30\lambda(2.5\phi) + 20\lambda(5\theta + 5\phi) = 40(\theta + \phi) + 40(\theta) + 60(2\phi)$$

$$+ 40(\theta + 2\phi) + 40(\theta + \phi)$$

$$\lambda(100\theta + 175\phi) = 160\theta + 280\phi$$

$$\lambda = 1.6. \tag{3.30}$$

This work equation could have been obtained by adding the two work equations (3.28) and (3.29), since the displacements and hinge rotations of these mechanisms were added to form the mechanism of Fig. 3.6(d). The same corresponding value of λ is found regardless of the relative magnitudes of θ and ϕ.

The actual collapse mechanism thus involves five plastic hinges and has two degrees of freedom. Since this frame has a number of redundancies $r = 3$, there are $(r + 2)$ plastic hinges in the collapse mechanism. A result of this kind only occurs at certain definite values of the ratios of the applied loads. Thus if the horizontal load were kept constant at the value 20λ kN, it is easily seen that if the vertical load was 29λ kN collapse would occur by sway, whereas if this load was 31λ kN collapse would occur by the combined mechanism.

3.5.3 Continuous beams

A continuous beam resting on several supports will usually collapse in a mechanism which is either partial or overcomplete. Consider as an illustration the continuous beam shown in Fig. 3.7(a), which rests on four simple supports and is of uniform section throughout, with plastic moment 10 kN m. For the loading shown, it is evident that collapse can only occur by one of the two mechanisms which are shown in Fig. 3.7(b) and (c).

The beam has two redundancies, and there must therefore be two equations of equilibrium connecting the bending moments at the four numbered cross sections. These equations may be derived by the virtual displacements method, and are as follows:

$$20\lambda = -M_1 + 2M_2 - M_3. \tag{3.31}$$

$$20\lambda = -M_3 + 2M_4. \tag{3.32}$$

If the mechanism of Fig. 3.7(c) is assumed to be the actual collapse mechanism, the bending moments at the plastic hinges are

$$M_3 = -10, \quad M_4 = +10.$$

Substituting these values in Equations (3.31) and (3.32) it is found that

$$\lambda = 1.5$$

$$-M_1 + 2M_2 = 20. \tag{3.33}$$

It is easy to see that this equation can be satisfied by pairs of values of M_1 and M_2 neither of which exceeds the plastic moment 10 kN m in magnitude, for example

$$M_1 = -6, \quad M_2 = +7.$$

It therefore follows from the uniqueness theorem that the mechanism of Fig. 3.7(c) is the actual collapse mechanism and that λ_c is 1.5. This is a case of partial collapse because there are only two plastic hinges involved in the collapse mechanism and the frame has two redundancies.

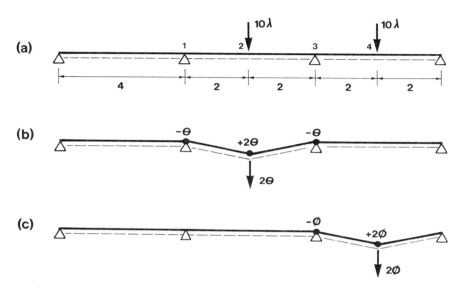

Fig. 3.7 *Continuous beam*

      (a) Dimensions and loading. Units: m, kN
      (b) Collapse of central span
      (c) Collapse of right-hand span

A case of overcomplete collapse would arise for this beam if the load on the right-hand span were reduced to 7.5 kN. It can easily be verified by a kinematical analysis that the value of λ corresponding to the mechanism of Fig. 3.7(c) is increased from 1.5 to 2, and this is also the value of λ which corresponds to the mechanism of Fig. 3.7(b). In this case a collapse mechanism with two degrees of freedom could be formed by adding the hinge rotations and displacements of both these mechanisms.

There is ample experimental evidence that the collapse loads for continuous beams can be predicted closely by the plastic theory. Maier-Leibnitz (1936) has given a critical review of the many tests carried out up to 1936 by himself, and by other investigators. In some of the early tests by Maier-Leibnitz (1928) intermediate supports were lowered before the test commenced, and tests of this kind were also carried out by Horne (1952a); the results showed that the collapse load was not thereby affected. Horne (1952a) also showed that the collapse load was not affected by non-proportional loading.

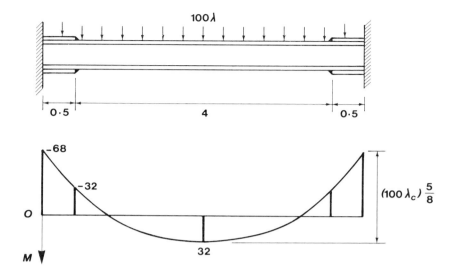

Fig. 3.8 *Fixed-ended beam with flange plates. Units:* m, kN

3.5.4 *Fixed-ended beam strengthened by flange plates*

A final example of overcomplete collapse is provided by the case of a fixed-ended beam which is strengthened by flange plates welded on at its ends, as illustrated in Fig. 3.8. Problems of this general type have been considered in detail by Horne (1952b) and also by Popov and Willis (1957). It will here be supposed that these plates each extend over a length 0.5 m, as shown, the span of the beam being 5 m. The beam, whose unstrengthened cross section has a plastic moment 32 kN m, is subjected to a uniformly distributed load 100λ kN. The optimum effect is achieved if at collapse the plastic moment of the strengthened section is developed at the built-in ends, and in addition the plastic moment of the unstrengthened section is developed at the centre of the beam and also at the sections where the strengthening begins. The corresponding bending moment diagram is shown in Fig. 3.8. A simple calculation then shows that the value of the plastic moment at the ends of the beam is $32 \times 2.125 = 68$ kN m, and from the figure it is then seen that the collapse load factor λ_c is given by

$$(100\lambda_c)\tfrac{5}{8} = 32 + 68$$

$$\lambda_c = 1.6.$$

In this situation there are two possible collapse mechanisms. One of these is with two end hinges and a central hinge, and the second has two hinges at the termination of the strengthening together with a central hinge. A collapse mechanism with two degrees of freedom could therefore be formed without difficulty.

References

Baker, J.F. (1949), 'The design of steel frames', *Struct. Engr*, **27**, 397.

Baker, J.F. and Heyman, J. (1950), 'Tests on miniature portal frames', *Struct. Engr*, **28**, 139.

Baker, J.F. and Roderick, J.W. (1952), 'Tests on full-scale portal frames', *Proc. Inst. Civil Engrs*, **1**, (part I), 71.

Feinberg, S.M. (1948), 'The principle of limiting stress' (Russian), *Prikladnaya Matematika i Mekhanika*, **12**, 63.

Greenberg, H.J. (1949), 'The principle of limiting stress for structures', 2nd Symposium on Plasticity, Brown Univ., April 1949.

Greenberg, H.J. and Prager, W. (1952), 'On limit design of beams and frames', *Trans. Am. Soc. Civil Engrs*, **117**, 447. (First published as Tech. Rep. A18-1, Brown Univ., 1949).

Gvozdev, A.A. (1936), 'The determination of the value of the collapse load for statically indeterminate systems undergoing plastic deformation', Proceedings of the Conference on Plastic Deformations, December 1936, Akademiia Nauk S.S.S.R., Moscow-Leningrad, 1938, 19, (tr., R.M. Haythornthwaite, *Int. J. Mech. Sci.*, **1**, 322, 1960).

Horne, M.R. (1950), 'Fundamental propositions in the plastic theory of structures', *J. Inst. Civil Engrs*, **34**, 174.

Horne, M.R. (1952a), 'Experimental investigations into the behaviour of continuous and fixed-ended beams', prelim. publ., 4th Congr. Int. Assoc. Bridge Struct. Eng., Cambridge, 1952, 147.

Horne, M.R. (1952b), 'Determination of the shape of fixed-ended beams for maximum economy according to the plastic theory', prelim. publ., 4th Congr. Int. Assoc. Bridge Struct. Eng., Cambridge, 1952, 111.

Kist, N.C. (1917), 'Leidt een Sterkteberekening, die Uitgaat van de Evenredigheid van Kracht en Vormverandering, tot een goede Constructie van Ijzeren Bruggen en gebouwen?', Inaugural Dissertation, Polytechnic Institute, Delft.

Maier-Leibnitz, H. (1928), 'Beitrag zur Frage der tatsächlichen Tragfähigkeit einfacher und durchlaufender Balkenträger aus Baustahl St. 37 und aus Holz', *Bautechnik*, **6**, 11.

Maier-Leibnitz, H. (1936), 'Versuche, Ausdeutung und Anwendung der Ergebnisse', prelim. publ., 2nd Congr. Int. Assoc. Bridge Struct. Eng., Berlin, 1936, 97.

Popov, E.P. and Willis, J.A. (1957), 'Plastic design of cover plated continuous beams', *J. Eng. Mech. Div., Proc. Am. Soc. Civil Engrs*, **84**, Paper 1495.

Schilling, C.G., Schutz, F.W. and Beedle, L.S. (1956), 'Behaviour of welded single-span frames under combined loading', *Weld. J., Easton, Pa.*, **35**, 234-s.

Examples

1. A uniform continuous beam whose plastic moment is 28 kN m rests on five simple supports A, B, C, D and E:

$$AB = 3\,m, \quad BC = CD = 4\,m, \quad DE = 5\,m.$$

Each span carries a concentrated load at its mid-point, these loads being:

$AB = 25\lambda \, kN, \quad BC = 25\lambda \, kN, \quad CD = 35\lambda \, kN, \quad DE = 12.5\lambda \, kN.$

Find the collapse load factor, and determine the limits between which the reaction at A must lie at collapse.

2. A uniform continuous beam whose plastic moment is M_p rests on three simple supports at A, B and C, where $AB = BC = l$. The span AB is unloaded, and BC carries a load W uniformly distributed over the span l. Show that at collapse the plastic hinge which forms in BC is located at a distance $(\sqrt{2} - 1)l$ from C, and find the value of W which would cause collapse. By means of an elastic analysis show that if W is steadily increased from zero yield first occurs at a distance $7l/16$ from C.

3. A continuous beam rests on four simple supports A, B, C and D:

$$AB = BC = CD = 3 \, m.$$

Each span carries a uniformly distributed load, as follows:

$$AB = 50 \, kN, \quad BC = 100 \, kN, \quad CD = 60 \, kN.$$

The beam is to be designed so that it is of uniform section in each span, but the plastic moments of the spans may all be different. Find the required value of the plastic moment for each span so that collapse would just occur with a load factor of unity. It may be assumed that the plastic moment of BC is greater than for AB or CD. Use the answer to example 2 for AB and CD, and use Equations (3.16) and (3.17) when considering BC.

4. A uniform fixed-ended beam of length l and plastic moment M_p is subjected to a uniformly distributed load W together with a concentrated load P at a distance $l/3$ from one end of the beam. Find the value of W which would cause collapse for the following three values of P: $0.25W, 0.5W$ and W.

5. For the fixed-ended beam of Fig. 3.8 the collapse load factor λ_c was 1.6 if the flange plates increased the plastic moment from 32 kN m to 68 kN m over a distance of 0.5 m at each end of the beam. If the ends of the beam were not strengthened in this way, find the central length which would require strengthening to the same increased value of the plastic moment to achieve the same value of λ_c.

6. Write down two equations of equilibrium for the frame of Fig. 3.6(a). Assuming collapse by the combined mechanism, verify that the corresponding value of λ is 1.6 and that the bending moment at section 2 is $+40$ kN m.

7. A fixed-base rectangular portal frame is of height l and span $2l$, and is of uniform section throughout, with plastic moment M_p. The frame carries a horizontal load H applied at the top of one of the columns, and also a vertical load V at the centre of the beam. Find the value of W which would cause collapse for the following pairs of values of H and V, and in each case determine the bending

moment distribution at collapse.

(a) $H = W$, $V = 0$.
(b) $H = W$, $V = 0.5W$.
(c) $H = W$, $V = W$.
(d) $H = W$, $V = 2W$.
(e) $H = W$, $V = 3W$.

Confirm that in cases (b) and (d) the collapse is overcomplete. For case (d) sketch the overcomplete collapse mechanism with two degrees of freedom, and show that the value of W_c can be found from the work equation corresponding to this mechanism.

8. In a pinned-base rectangular portal frame ABCD the columns AB and CD are each of height 3 m and the beam BC is length 9 m. The members are of uniform section throughout, with plastic moment 30 kN m. The frame carries a horizontal load 10λ kN applied at C, and also a vertical load 10λ kN at a distance 3 m from B. Find the collapse load factor when the horizontal load acts in the direction BC, and also when it acts in the direction CB.

9. Perform a statical analysis for the frame of Fig. 3.3(a), assuming the sway mechanism of collapse. Using Equations (3.16) and (3.17), find the position and magnitude of the greatest bending moment occurring in the beam, and hence determine a lower bound on the value of λ_c.

10. For the frame of Fig. 3.3(a), find the value of λ corresponding to the combined mechanism, with the hinge in the beam located at the position of maximum bending moment found in example 9. Using Equations (3.16) and (3.17), find the magnitude of the greatest bending moment in the beam in the corresponding bending moment distribution, and hence determine a lower bound on the value of λ_c.

11. A fixed-base rectangular portal frame is of height and span 5 m. The columns each have a plastic moment 24 kN m and the beam has a plastic moment 12 kN m.

One of the columns is subjected to a uniformly distributed horizontal load 15λ kN. Show that the collapse mechanism is the sway mechanism, with plastic hinges at the two bases and the two beam/column joints, and determine the value of the collapse load factor. Investigate the bending moment distribution in the loaded column using Equations (3.16) and (3.17). What would be the value of the collapse load factor if the plastic moments of the columns were reduced to 12 kN m?

12. In a fixed-base rectangular portal frame ABCD the columns AB and CD are of height 4 m and 6 m, respectively. The base D is lower than the base A by a height 2 m, so that the beam BC, of length 4 m, is horizontal. All the members of the frame have the same plastic moment 20 kN m.

The beam BC carries a central concentrated vertical load 20λ kN, and a concentrated horizontal load 8λ kN is applied at C in the direction BC. Find the value of λ at which collapse would occur. Show also that if the horizontal load is reversed in direction, the collapse load factor is changed, and find its new value.

13. A uniform beam of length l and plastic moment M_p is simply supported at one end and rigidly built-in at the other end. A concentrated load W may be applied anywhere within the span. Find the smallest value of M_p such that collapse would just occur when the load was in its most unfavourable position.

4 Methods of Plastic Design

4.1 Introduction

In this chapter two methods for designing a frame so that plastic collapse would just occur at a given load factor are described. It is assumed that failure by buckling does not occur before the plastic collapse load is attained; the problems associated with the buckling of members which have entered the plastic range are beyond the scope of this book.

The first method to be described is the *trial-and-error* method. This method is suitable only when, on the basis of previous experience, the collapse mechanism is believed to be known. It consists essentially of verifying that, for the assumed mechanism of collapse, a safe and statically admissible bending moment distribution can be found.

When the collapse mechanism is not known, the method of *combining mechanisms* may be used. The procedure consists of investigating a number of possible collapse mechanisms which are formed by combining certain *independent mechanisms*. When it is thought that the correct collapse mechanism has been found, the result is checked by the same procedure as in the trial-and-error method. Other methods, based on the use of linear programming techniques, have been developed. These are beyond the scope of this book, but are discussed briefly in a concluding section.

4.2 Trial-and-error method

The trial-and-error method will be illustrated by its application to the pitched-roof portal frame shown in Fig. 4.1(a). The loads shown are due to dead and superimposed vertical loading, together with the effect of wind. All the loads are uniformly distributed along the members; for convenience they are indicated by broken arrows which show the resultants acting at the centres of the members.

All the joints, including those at the bases, are assumed to be capable of developing the plastic moment. The members all have the same plastic moment M_p, and it is required to find the value of M_p such that collapse would just occur at a load factor λ of 1.6.

Since the method consists essentially of a statical check of an assumed mechanism, the first step is to develop the equations of equilibrium, using the method of virtual displacements. Attention will initially be confined to the bending moments at the ends and centres of the four members, these positions being numbered from 1–9 in Fig. 4.1(a). Since the frame has three redundancies, there must be six independent equations of equilibrium relating the nine unknown bending moments.

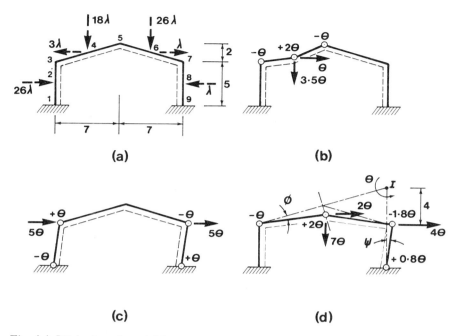

(a)

(b)

(c)

(d)

Fig. 4.1 *Pitched-roof portal frame*
  (a) Dimensions and loading: all loads uniformly distributed. Units: m, kN
  (b) Beam-type mechanism
  (c) and (d) Sway mechanisms

Four of these equations can be derived from beam-type mechanisms, one for each member. A typical mechanism of this kind, for the left-hand rafter, is shown in Fig. 4.1(b). Its motion is defined by a clockwise rotation θ of 34 about the joint 3. Since the mid-point is at a horizontal distance 3.5 m from 3, its vertical movement is 3.5θ m downwards. Its height above the left-hand end is 1.0 m, and the horizontal movement is therefore 1.0θ m to the right, as shown.

The two other mechanisms which will be used are shown in Fig. 4.1(c) and (d). The sway mechanism of Fig. 4.1(c) is of a kind encountered previously, and calls for no comment, but the other sway mechanism of Fig. 4.1(d) is more complex. In this mechanism the left-hand rafter 35 rotates about the joint 3, so that the apex 5 moves in a direction perpendicular to 35. The column 79 rotates about its base 9, and the joint 7 therefore moves horizontally. This defines the

directions in which the ends 5 and 7 of the right-hand rafter 57 move; conse-
quently this member rotates about the *instantaneous centre* I shown in Fig.
4.1(d). The motion of the mechanism is specified by a counterclockwise rotation
θ of the rafter 57 about I.

The horizontal movement of joint 7 is seen from the figure to be 4θ m. If the
clockwise rotation of the column 79 is specified as ψ, it is seen that $4\theta = 5\psi$, so
that $\psi = 0.8\theta$. The clockwise rotation ϕ of the rafter 35 can by a similar argument be shown to be θ.

Since the rotation of the rafter 57 is θ counterclockwise and that of the column
79 is 0.8θ clockwise, it follows that the rotation at the hinge at joint 7 is 1.8θ in
magnitude; by inspection its sign is negative. The rotation at the hinge at the
apex 5 can similarly be shown to be $+2\theta$. This completes the kinematical
analysis of this mechanism.

The six equations of equilibrium derived from the six mechanisms are as
follows:

$$32.5\lambda = -M_1 + 2M_2 - M_3. \quad \text{\small beam mechanism} \tag{4.1}$$

$$30\lambda = -M_3 + 2M_4 - M_5. \quad \text{\small beam} \tag{4.2}$$

$$45\lambda = -M_5 + 2M_6 - M_7. \quad \text{\small beam} \tag{4.3}$$

$$1.25\lambda = -M_7 + 2M_8 - M_9. \quad \text{\small beam} \tag{4.4}$$

$$52.5\lambda = -M_1 + M_3 - M_7 + M_9. \quad \text{\small sway} \tag{4.5}$$

$$152\lambda = -M_3 + 2M_5 - 1.8M_7 + 0.8M_9. \quad \text{\small gable} \tag{4.6}$$

The calculations will be carried out by assuming an arbitrary value for M_p,
40 kN m, and treating λ as the parameter to be determined in the first instance,
as in the previous chapter. The correct value of M_p to ensure a collapse load factor of 1.6 is found as a final step by simple proportions.

It is known that for this type of frame and loading the collapse mechanism is
the mechanism of Fig. 4.1(d), subject only to minor adjustments in hinge positions arising from the fact that the loads on the members are uniformly distributed. It is therefore assumed that

$$M_3 = -40, \quad M_5 = +40, \quad M_7 = -40, \quad M_9 = +40 \tag{4.7}$$

the units (kN m) being omitted for convenience.

Substitution in Equation (4.6) gives the value of λ, and the other five
equations are easily solved in turn to give:

$$\lambda = 1.474$$

$$M_1 = -37.4, \quad M_2 = -14.7, \quad M_4 = +22.1,$$

$$M_6 = +33.2, \quad M_8 = +0.9. \tag{4.8}$$

None of these bending moments exceeds the plastic moment 40 in magnitude. The mechanism of Fig. 4.1(d) is therefore the actual collapse mechanism, subject only to the reservation that since each member carries uniformly distributed loads it is possible that a bending moment greater in magnitude than the plastic moment might occur at a section other than those so far examined.

For this situation, the complete bending moment diagram for the frame is shown in Fig. 4.2, in which the frame has been opened out to form a horizontal datum. It will be seen that the form of the bending moment distribution in both the columns is such that the plastic moment is not exceeded. However, for each of the rafters this possibility exists. Using the method explained in Chapter 3, Equations (3.16) and (3.17), it can be shown that there are maximum bending moments in the two rafters, the largest of these bending moments being $+45.2$ in the right-hand rafter, occurring 1.44 m from the apex.

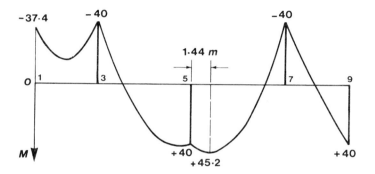

Fig. 4.2 *Bending moment diagram for pitched-roof portal frame*

A fresh analysis is now made assuming that a plastic hinge occurs at this position of maximum bending moment, rather than at the apex; the other three plastic hinges are left at sections 3, 7 and 9. The bending moment at this position is defined as M_{10}, and a further equation of equilibrium is required which involves this moment. This can be derived from the beam-type mechanism shown in Fig. 4.3, for which

$$5.61\psi = 1.39\theta$$

$$\psi = 0.25\theta.$$

The corresponding equation is:

$$17.85\lambda = -M_5 + 1.25M_{10} - 0.25M_7. \tag{4.9}$$

The plastic moments in the new mechanism are:

$$M_3 = -40, \quad M_{10} = +40, \quad M_7 = -40, \quad M_9 = +40. \tag{4.10}$$

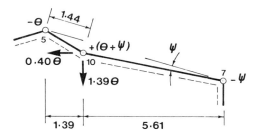

Fig. 4.3 *Beam-type mechanism for rafter*

Substituting in Equations (4.1)–(4.6) and (4.9) gives

$$\lambda = 1.404$$

$$M_1 = -33.7, \quad M_2 = -14.1, \quad M_4 = +18.4, \quad M_5 = +34.7,$$

$$M_6 = +29.0, \quad M_8 = +0.9. \tag{4.11}$$

The bending moment distribution is similar to that of Fig. 4.2, and need not be plotted. The maximum bending moment in the left-hand rafter is now reduced to $+35.0$. In the right-hand rafter the maximum moment now occurs at a position 1.49 m from the apex, or 0.05 m from section 10, and is $+40.00$ to four significant figures.

Thus, with the plastic moments specified in equations (4.10) and (4.11), the resulting bending moment distribution is both safe and statically admissible with a load factor of 1.404. By the uniqueness theorem this is therefore the collapse load factor for the frame when $M_p = 40 \, \text{kN m}$. If all the loads and bending moments were increased in the ratio $1.6/1.404$, the bending moment distribution would remain safe, provided that the plastic moment was increased to

$$M_p = \left(\frac{1.6}{1.404}\right) 40 = 45.6 \, \text{kN m},$$

and would also be statically admissible with the load factor 1.6. The required value of M_p is therefore 45.6 kN m.

The trial-and-error method is especially suited to single-bay portal frames, for which the collapse mechanism is generally known from previous experience. It is often carried out by a semi-graphical method, in which the bending moment diagram is represented as the difference between the free and reactant diagrams.

A comprehensive account of the method for portal frames has been given by Heyman (1957). The design of single-bay portal frames has also been discussed by several authors, including Harrison (1960). The analysis of a three-bay portal frame by this method was described by Baker (1949), and it was used by Hendry (1955) to analyse Vierendeel girders. Various tests on portal frames have been reported, for instance the full-scale tests of Baker and Eickhoff (1955, 1956) and Driscoll and Beedle (1957).

Difficulties arise in the trial-and-error method when the actual collapse mechanism could be of the partial type, so that with r redundancies there may be fewer than $(r + 1)$ plastic hinges at collapse. When investigating an assumed partial collapse mechanism there is no difficulty in determining the statically admissible bending moment distribution in that part of the frame which is statically determinate at collapse, and establishing whether this distribution is also safe. However, it is also necessary to examine the remainder of the frame, in which the bending moment distribution is not uniquely determined, to see whether among all the possible statically admissible bending moment distributions at least one safe distribution exists. Such an investigation may present considerable difficulty, particularly if this remaining portion of the frame is highly redundant.

Experience in the design of similar structures may afford the necessary guidance as to the likelihood of an assumed collapse mechanism being the correct one. However, in facing unusual problems there is a need for a method by which a close approximation to the actual collapse mechanism can be found quickly, even though this mechanism may be of the partial type. The method of combining mechanisms, which will now be described, enables this to be done.

4.3 Method of combining mechanisms

The essential notion underlying this method, which was developed by Neal and Symonds (1952a, 1952b), is that for a given frame and loading every possible collapse mechanism can be regarded as a combination of a certain number of *independent mechanisms*. For each possible collapse mechanism a work equation can be written down from which the corresponding value of the load factor λ is found. The actual collapse mechanism is distinguished from among all the possible mechanisms by the fact that it has the lowest corresponding value of λ, by the kinematic theorem. The independent mechanisms with low corresponding values of λ are therefore examined to see whether they can be combined to form a mechanism which gives an even lower value of λ. It is only necessary to examine a few of the more likely combinations in order to arrive at a mechanism which is almost certainly the actual collapse mechanism. A statical check is then performed to verify the result. The basic procedure is first explained with reference to a simple rectangular frame problem, and this is followed by some more complex examples.

4.3.1 *Rectangular frame*

In the rectangular frame whose dimensions and loading are shown in Fig. 4.4(a), the plastic moment of each column is to be 50 per cent greater than that of the beam. It is required to find the values of these plastic moments which would provide a load factor of 1.5 against plastic collapse. The initially assumed values of 45 kN m and 30 kN m, shown in the figure, will be adjusted at the end of the analysis.

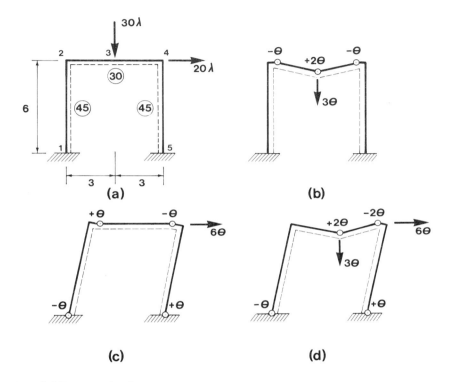

Fig. 4.4 *Rectangular frame*

    (a) Dimensions and loading. Units: m, kN. Plastic moments shown thus: (M_p)
    (b) Beam mechanism
    (c) Sway mechanism
    (d) Combined mechanism

The number n of possible plastic hinge positions is 5; these positions are numbered in the figure. The number of redundancies r is 3, and so there are $(n - r) = 2$ independent equations of equilibrium connecting the bending moments at the five numbered cross sections. These equations, which are required for the statical check, are derived by the method of virtual displacements from the two mechanisms shown in Fig. 4.4(b) and (c), and are:

$$90\lambda\theta = M_2(-\theta) + M_3(+2\theta) + M_4(-\theta)$$

$$90\lambda = -M_2 + 2M_3 - M_4. \qquad (4.12)$$

$$120\lambda\theta = M_1(-\theta) + M_2(+\theta) + M_4(-\theta) + M_5(+\theta)$$

$$120\lambda = -M_1 + M_2 - M_4 + M_5. \tag{4.13}$$

The beam mechanism of Fig. 4.4(b) is now considered as a possible collapse mechanism. The hinges are treated as plastic hinges, at which the work absorbed is always positive, regardless of the sense of the rotation. At sections 2 and 4 the hinges are located in the beam, which has a smaller plastic moment than the columns. The work equation is

$$90\lambda\theta = 30(\theta) + 30(2\theta) + 30(\theta) = 120\theta$$

$$\lambda = 1.333. \tag{4.14}$$

Similarly, for the sway mechanism of Fig. 4.4(c), the work equation is

$$120\lambda\theta = 45(\theta) + 30(\theta) + 30(\theta) + 45(\theta) = 150\theta$$

$$\lambda = 1.25. \tag{4.15}$$

Now consider the mechanism of Fig. 4.4(d). This combined mechanism is obtained by adding the displacements and hinge rotations of the beam and sway mechanisms. The work equation for this mechanism will not be derived directly, but will be deduced from the work equations (4.14) and (4.15) for the two mechanisms which have been combined.

Examination of Fig. 4.4(b), (c) and (d) reveals that the displacements in the combined mechanism are obtained by adding the displacements in the beam and sway mechanisms. The work done by the loads in the combined mechanism is therefore the sum of the work done in the beam and sway mechanisms.

The hinge rotations in the combined mechanism are also obtained by adding the hinge rotations in the two mechanisms combined. However, the work absorbed at the hinges is not additive in each case, as can be seen from Table 4.1.

Table 4.1 *Hinge rotations and work absorbed at plastic hinges*

Section	Beam mechanism		Sway mechanism		Combined mechanism	
	Hinge rotation	Work absorbed	Hinge rotation	Work absorbed	Hinge rotation	Work absorbed
1	–	–	$-\theta$	45θ	$-\theta$	45θ
2	$-\theta$	30θ	$+\theta$	30θ	–	–
3	$+2\theta$	60θ	–	–	$+2\theta$	60θ
4	$-\theta$	30θ	$-\theta$	30θ	-2θ	60θ
5	–	–	$+\theta$	45θ	$+\theta$	45θ

At section 2, the work absorbed is 30θ for both the beam and sway mechanisms. However, the hinge rotations $-\theta$ and $+\theta$ combine to give zero rotation and therefore zero work absorbed in the combined mechanism. At the other four cross sections, the work absorbed is additive. The work equation for the combined mechanism is therefore derived as follows:

$$\text{Beam:} \quad 90\lambda\theta = 120\theta; \quad \lambda = 1.333. \tag{4.14}$$

$$\text{Sway:} \quad \underline{120\lambda\theta = 150\theta}; \quad \lambda = 1.25. \tag{4.15}$$

$$\text{Combined:} \quad 210\lambda\theta = 270\theta - 2 \times 30\theta$$

$$= 210\theta$$

$$\lambda = 1. \tag{4.16}$$

Because of the cancellation of the hinge at section 2, and the consequential reduction in the work absorbed, the value of λ for the combined mechanism is less than for each of the mechanisms combined. Since in this simple example there is only one possible combination of mechanisms, it is concluded that the combined mechanism is the actual collapse mechanism.

The solution is readily checked by statics. The plastic moments in the collapse mechanism are:

$$M_1 = -45, \quad M_3 = +30, \quad M_4 = -30, \quad M_5 = +45.$$

Substituting in Equations (4.12) and (4.13), it is found that $M_2 = 0$ and $\lambda = 1$, and this confirms the analysis.

Since a load factor of 1.5 is required, the plastic moments initially assumed must be increased by the factor 1.5, so becoming $30 \times 1.5 = 45 \text{ kN m}$ for the beam and $45 \times 1.5 = 67.5 \text{ kN m}$ for the columns.

The correspondence between a plastic collapse mechanism and the breaking down of an equation of equilibrium was emphasized in Chapter 3. The combining mechanisms method of analysis is perhaps best understood on this basis. The two independent equations of equilibrium corresponding to the beam and sway mechansims are

$$90\lambda\theta = M_2(-\theta) + M_3(+2\theta) + M_4(-\theta). \tag{4.12}$$

$$120\lambda\theta = M_1(-\theta) + M_2(+\theta) + M_4(-\theta) + M_5(+\theta). \tag{4.13}$$

Adding these two equations, it is found that

$$210\lambda\theta = M_1(-\theta) + M_3(+2\theta) + M_4(-2\theta) + M_5(+\theta). \tag{4.17}$$

This is precisely the equation which results from applying the method of virtual displacements to the combined mechanism of Fig. 4.4(d). It is not, of course, independent of Equations (4.12) and (4.13).

Since no bending moment can exceed the plastic moment in magnitude, it follows that M_1 and M_5 must lie between the limits ± 45, while M_2, M_3 and M_4 must lie between the limits ± 30. The breakdown of the above three equations is therefore expressed as follows:

$$90\lambda\theta = (-30)(-\theta) + 30(+2\theta) - 30(-\theta) = 120\theta$$
$$120\lambda\theta = (-45)(-\theta) + 30(+\theta) - 30(-\theta) + 45(+\theta) = 150\theta$$
$$210\lambda\theta = -45(-\theta) + 30(+2\theta) - 30(-2\theta) + 45(+\theta) = 210\theta,$$

and these three equations are the work equations (4.14), (4.15) and (4.16), for the beam, sway and combined mechanisms, respectively. The combination of the beam and sway mechanisms has its exact counterpart in the combining of the two equations of equilibrium, (4.12) and (4.13), to form Equation (4.17). In particular, the hinge cancellation at section 2 corresponds to the elimination of M_2 between Equations (4.12) and (4.13) which results from their addition.

This example serves to illustrate the fact that in general there will be a number of independent mechanisms which is equal to the number of independent equations of equilibrium. If there are n possible plastic hinge positions and r redundancies, there will be $(n-r)$ independent equations of equilibrium and therefore $(n-r)$ independent mechanisms. The essence of the combining mechanisms technique is to identify these independent mechanisms and to explore those combinations in which hinge cancellations are likely to produce load factors which are lower than the lowest load factor arising from any of the independent mechanisms.

It will be appreciated that in the example of Fig. 4.4 any pair of the three mechanisms shown in this figure could have been selected as the two independent mechanisms. The reason for choosing the beam and sway mechanisms as the two independent mechanisms is that if the combined mechanism were selected as one of the independent mechanisms, the combination of the independent mechanisms would involve the subtraction of hinge rotations and displacements. For example, choosing the combined mechanism and the beam mechanism as the two independent mechanisms, the sway mechanism is derived by subtracting the displacements and hinge rotations of the beam mechanism from those of the combined mechanism. This would lead to some awkwardness in the corresponding calculations.

4.3.2 *Two-bay rectangular frame*

The technique of combining mechanisms will now be applied to the frame whose dimensions and loading are shown in Fig. 4.5(a). All the members of this frame are to have the same plastic moment, and there is to be a load factor of 1.4 against plastic collapse. The plastic moment is initially assumed to be 30 kN m.

The ten possible plastic hinge positions are numbered in the figure, and the frame has six redundancies. The number of independent mechanisms is therefore

$$n - r = 10 - 6 = 4.$$

Three of these mechanisms are readily identified as the sway mechanism of Fig. 4.5(b) and the two beam mechanisms of Fig. 4.5(c) and (d). The fourth in-

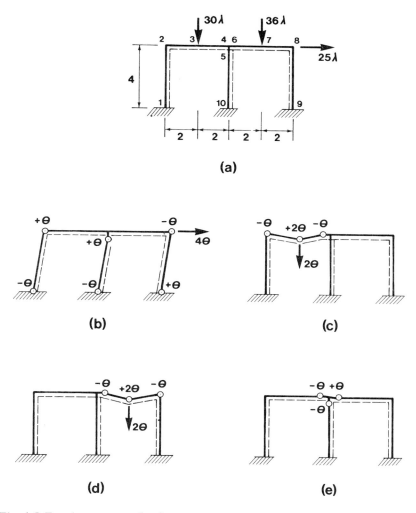

Fig. 4.5 *Two-bay rectangular frame*

    (a) Dimensions and loading. Units: m, kN
    (b) Sway mechanism
    (c) and (d) Beam mechanisms
    (e) Joint rotation mechanism

dependent mechanism is the joint rotation of Fig. 4.5(e). This mechanism is not, of course, a possible plastic collapse mechanism, as there is no applied couple at the central joint. However, it is clearly independent of the other three mechanisms, and can be combined with them to form possible mechanisms of collapse, as will be seen.

Equations of equilibrium can be derived from the four independent mechanisms by the method of virtual displacements. These equations are:

$$100\lambda\theta = M_1(-\theta) + M_2(+\theta) + M_5(+\theta) + M_{10}(-\theta) + M_8(-\theta) + M_9(+\theta)$$

$$60\lambda\theta = M_2(-\theta) + M_3(+2\theta) + M_4(-\theta)$$

$$72\lambda\theta = M_6(-\theta) + M_7(+2\theta) + M_8(-\theta)$$

$$0 = M_4(-\theta) + M_5(-\theta) + M_6(+\theta).$$

These equations reduce to

$$100\lambda = -M_1 + M_2 - M_{10} + M_5 + M_9 - M_8. \tag{4.18}$$

$$60\lambda = -M_2 + 2M_3 - M_4. \tag{4.19}$$

$$72\lambda = -M_6 + 2M_7 - M_8. \tag{4.20}$$

$$0 = -M_4 - M_5 + M_6. \tag{4.21}$$

If the sway and beam mechanisms are now viewed as possible plastic collapse mechanisms, their work equations can be derived immediately from the above equations. In each case, the work done is the same. At each plastic hinge the plastic moment is 30 and the work absorbed is always positive. The work equations are thus

Sway Fig. 4.5(b) $100\lambda\theta = 180\theta;$ $\lambda = 1.8$

Beam Fig. 4.5(c) $60\lambda\theta = 120\theta;$ $\lambda = 2$

Beam Fig. 4.5(d) $72\lambda\theta = 120\theta;$ $\lambda = 1.667.$

The three load factors obtained are not widely different. It is natural to begin by investigating the combination of the two mechanisms with the lowest load factors, these being the right-hand beam and sway mechanisms. Adding their displacements and hinge rotations gives rise to no hinge cancellation, as in Fig. 4.6(a). However, if the joint rotation mechanism of Fig. 4.5(e) is now added, the plastic hinge rotations $+\theta$ and $-\theta$ at sections 5 and 6 are cancelled, while a hinge rotation $-\theta$ appears at section 4. The effect is thus to reduce the work absorbed at the central joint from 60θ to 30θ. The resulting mechanism is shown in Fig. 4.6(b).

The work equation for this combination is derived as follows:

Sway Fig. 4.5(b) $100\lambda\theta = 180\theta;$ $\lambda = 1.8$

Right-hand beam Fig. 4.5(d) $\underline{72\lambda\theta = 120\theta};$ $\lambda = 1.667$

Combination Fig. 4.6(b) $172\lambda\theta = 300\theta - 30\theta = 270\theta$

$$\lambda = 1.570.$$

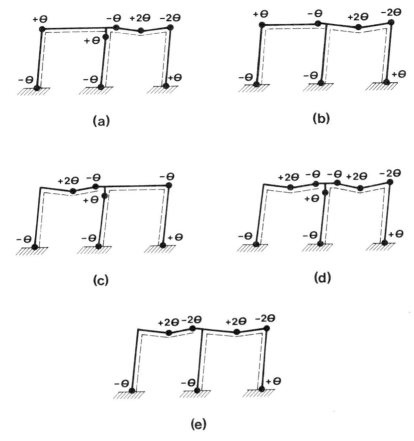

Fig. 4.6 *Two-bay rectangular frame: combinations of mechanisms*

This value of λ is lower than the value for either of the two mechanisms combined, because of the hinge cancellation. The calculation illustrates the role played by the joint mechanism. For any combination, the joint is rotated into the position which minimizes the work absorbed in the plastic hinges at the central joint. In this way, the joint mechanism features as an independent mechanism, although in itself it is not a possible collapse mechanism.

Another possible combination is obtained by adding the left-hand beam and sway mechanisms, as illustrated in Fig. 4.6(c). This results in the cancellation of the hinge at section 2. In this case there is no advantage to be gained by rotating the central joint. For each of the mechanisms combined the total work absorbed in the plastic hinges included a term 30θ for the hinge at section 2, so that a reduction of 60θ in the total work absorbed is achieved by this combination. The work equation is therefore derived as follows:

$$\text{Sway Fig. 4.5(b)} \quad 100\lambda\theta = 180\theta; \quad \lambda = 1.8$$

$$\text{Left-hand beam Fig. 4.5(c)} \quad \underline{60\lambda\theta = 120\theta}; \quad \lambda = 2$$

$$\text{Combination Fig. 4.6(c)} \quad 160\lambda\theta = 300\theta - 60\theta = 240\theta$$

$$\lambda = 1.5.$$

This is the lowest value of λ obtained so far. The only other possible combination is obtained by adding the right-hand beam mechanism to the mechanism just derived, Fig. 4.6(c). A direct addition gives the mechanism shown in Fig. 4.6(d). However, the effect of adding the joint rotation mechanism is shown in Fig. 4.6(e), and it will be seen that the work absorbed at the central joint is thereby reduced from 90θ to 60θ. The work equation for this mechanism is therefore obtained as follows:

$$\text{Combination Fig. 4.6(c)} \quad 160\lambda\theta = 240\theta; \quad \lambda = 1.5$$

$$\text{Right-hand beam Fig. 4.5(d)} \quad \underline{72\lambda\theta = 120\theta}; \quad \lambda = 1.667$$

$$\text{Combination Fig. 4.6(e)} \quad 232\lambda\theta = 360\theta - 30\theta = 330\theta$$

$$\lambda = 1.422.$$

This is the lowest load factor for all the mechanisms considered. The mechanism of Fig. 4.6(e) is therefore presumed to be the actual collapse mechanism. This conclusion will now be confirmed by a statical analysis.

The bending moments at the plastic hinges in the mechanism of Fig. 4.6(e) are as follows:

$$M_1 = -30, \quad M_3 = +30, \quad M_4 = -30, \quad M_7 = +30,$$

$$M_8 = -30, \quad M_9 = +30, \quad M_{10} = -30.$$

These values are substituted in the equations of equilibrium, (4.18)–(4.21), leaving four unknowns, namely M_2, M_5, M_6 and the value of λ. The solution is

$$M_2 = +4.66, \quad M_5 = +17.59, \quad M_6 = -12.41$$

$$\lambda = 1.422.$$

This value of λ agrees with the value found by the combining mechanisms analysis. Moreover, the values found for the three remaining bending moments are all less than the plastic moment 30 in magnitude. The bending moment distribution corresponding to the mechanism of Fig. 4.6(e) is thus both safe and statically admissible with $\lambda = 1.422$, confirming that this is the actual collapse mechanism.

Finally, it is noted that since the load factor is 1.422 if the plastic moment is 30 kN m, the required plastic moment for a load factor of 1.4 is

$$\left(\frac{1.4}{1.422}\right)30 = 29.5 \text{ kN m}$$

4.3.3 *Partial collapse*

In the problem just considered, the statical check was easy to perform because the collapse mechanism was complete; the number of redundancies r for the frame was six and there were seven hinges in the collapse mechanism. When the collapse mechanism is partial, with fewer than $(r + 1)$ hinges, the statical check is intrinsically more difficult, but the results of a combining mechanisms analysis are extremely useful as a guide.

To illustrate the procedure, consider the problem of Fig. 4.5(a), modified by changing the vertical load 36λ on the right-hand beam to 48λ. The other two loads and the frame dimensions remain the same, and all the members still have the same plastic moment, taken initially to be 30 kN m.

The combining mechanisms analysis need not be repeated here; the results are:

$$\text{Sway Fig. 4.5(b)} \quad 100\lambda\theta = 180\theta; \quad \lambda = 1.8$$

$$\text{Left-hand beam Fig. 4.5(c)} \quad 60\lambda\theta = 120\theta; \quad \lambda = 2$$

$$\text{Right-hand beam Fig. 4.5(d)} \quad 96\lambda\theta = 120\theta; \quad \lambda = 1.25$$

$$\text{Combination Fig. 4.6(b)} \quad 196\lambda\theta = 270\theta; \quad \lambda = 1.378$$

$$\text{Combination Fig. 4.6(c)} \quad 160\lambda\theta = 240\theta; \quad \lambda = 1.5$$

$$\text{Combination Fig. 4.6(e)} \quad 256\lambda\theta = 330\theta; \quad \lambda = 1.289$$

The lowest load factor is now for the right-hand beam mechanism of Fig. 4.5(d), which only involves three hinges, and so it is concluded that this is the actual collapse mechanism. To verify this result, the four independent equations of equilibrium are derived as before; they are

$$100\lambda = -M_1 + M_2 - M_{10} + M_5 + M_9 - M_8. \tag{4.22}$$

$$60\lambda = -M_2 + 2M_3 - M_4. \tag{4.23}$$

$$96\lambda = -M_6 + 2M_7 - M_8. \tag{4.24}$$

$$0 = -M_4 - M_5 + M_6. \tag{4.25}$$

The bending moments at the plastic hinges in the collapse mechanism are

$$M_6 = -30, \quad M_7 = +30, \quad M_8 = -30. \tag{4.26}$$

When these values are substituted in Equation (4.24), which was derived from the right-hand beam mechanism, it is found at once that $\lambda = 1.25$. However, the other three equations cannot be solved to determine the values of any of the remaining seven unknown bending moments.

In order to show that the correct collapse mechanism has been identified, a set of bending moments which is both safe and statically admissible with

$\lambda = 1.25$ must be established. This set must include the three values given in Equations (4.26). Guidance is provided by noting that the mechanism with the next lowest value of λ is the combination of Fig. 4.6(e), for which the corresponding value of λ is 1.289. This mechanism was obtained by adding the displacements and hinge rotations of the four independent mechanisms. The statical equivalent is to add the four equations of equilibrium, (4.22)–(4.25), giving

$$256\lambda = -M_1 + 2M_3 - 2M_4 + 2M_7 - 2M_8 + M_9 - M_{10}. \qquad (4.27)$$

Disregarding any other requirements of equilibrium, it is seen that since no bending moment can exceed the plastic moment in magnitude, this equation breaks down when the seven bending moments involved have the values

$$M_1 = -30, \quad M_3 = +30, \quad M_4 = -30, \quad M_7 = +30,$$
$$M_8 = -30, \quad M_9 = +30, \quad M_{10} = -30. \qquad (4.28)$$

With these values, Equation (4.27) becomes

$$256\lambda = 330; \qquad \lambda = 1.289.$$

Equation (4.27) is therefore close to breaking down when $\lambda = 1.25$. It follows that the set of bending moments which is sought cannot involve values differing much from those in Equations (4.28). Since complete collapse requires seven plastic hinges, and there are only three in the actual collapse mechanism, four values must be selected to enable the equations of equilibrium to be solved.

Two of the values appearing in Equations (4.28), namely $M_7 = +30$ and $M_8 = -30$, occur in the actual collapse mechanism. An arbitrary choice of four of the remaining five is

$$M_1 = -30, \quad M_3 = +30, \quad M_4 = -30, \quad M_9 = +30. \qquad (4.29)$$

Substituting these values in the equations of equilibrium, with $\lambda = 1.25$, it is found that

$$M_2 = +15, \quad M_5 = 0, \quad M_{10} = -20. \qquad (4.30)$$

The set of bending moments contained in Equations (4.29) and (4.30), together with the three plastic moments involved in the collapse mechanism, Equations (4.26), are statically admissible with $\lambda = 1.25$ and also safe, since none of the values exceeds $30\,\text{kN}\,\text{m}$ in magnitude. This confirms that the collapse mechanism is the right-hand beam mechanism, and that the collapse load factor is 1.25. To provide a load factor of 1.4, the required value of the plastic moment would therefore be

$$\left(\frac{1.4}{1.25}\right)30 = 33.6\,\text{kN}\,\text{m}.$$

4.3.4 *Distributed loads*

The frame whose dimensions and loading are shown in Fig. 4.7(a) will now be analysed to illustrate the technique for dealing with distributed loads. Each member carries a uniformly distributed load, whose resultant is indicated by the broken arrows. Trial values of the plastic moments are shown against each member; these are in the ratios $1:2:3$ for the top rectangular frame, the bottom columns and the bottom beam, respectively. The load factor against plastic collapse is to be 1.5.

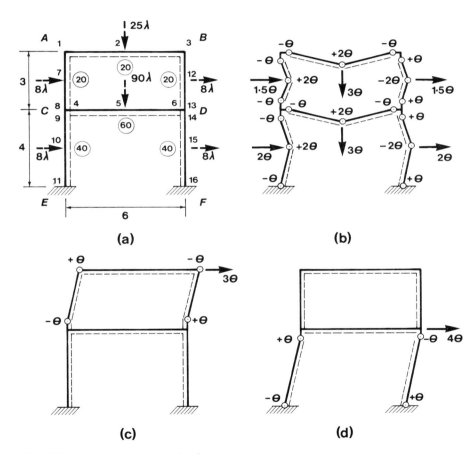

(a)   (b)

(c)   (d)

Fig. 4.7 *Two-storey rectangular frame*

   (a) Dimensions and loading: all loads uniformly distributed. Units: m, kN. Plastic
      moments shown thus: $\left(M_{\mathbf{p}}\right)$

   (b) Beam mechanisms
   (c) and (d) Sway mechanisms

In the first stage of the analysis it is assumed that plastic hinges can only occur at the ends and mid-points of members. When the correct collapse mechanism, subject only to this restriction, has been deduced, appropriate adjustments to the hinge positions will be made by the method described in Section 3.4.

On this basis there are 16 possible plastic hinge positions, at the cross sections numbered in the figure. Since the frame has six redundancies, the number of independent mechanisms is

$$n - r = 16 - 6 = 10.$$

The independent mechanisms can be identified as follows. There is a beam-type mechanism for each of the six members, two beams and four columns. These six mechanisms are shown for convenience on the same diagram, Fig. 4.7(b). There are also two independent sway mechanisms, one for each storey, as shown in Fig. 4.7(c) and (d). Finally, there are two joint rotation mechanisms, one at joint C and the other at D. The ten independent mechanisms are therefore made up as follows:

6 beam-type

2 sway

2 joint rotation

$\overline{10}$ independent mechanisms.

The independent equations of equilibrium derived from these mechanisms are:

$$
\begin{aligned}
37.5\lambda &= -M_1 + 2M_2 - M_3 \\
135\lambda &= -M_4 + 2M_5 - M_6 \\
6\lambda &= -M_1 + 2M_7 - M_8 \\
8\lambda &= -M_9 + 2M_{10} - M_{11} \\
6\lambda &= +M_3 - 2M_{12} + M_{13} \\
8\lambda &= +M_{14} - 2M_{15} + M_{16} \\
24\lambda &= -M_8 + M_1 - M_3 + M_{13} \\
96\lambda &= -M_{11} + M_9 - M_{14} + M_{16} \\
0 &= +M_4 + M_8 - M_9 \\
0 &= -M_6 - M_{13} + M_{14}.
\end{aligned}
\tag{4.31}
$$

The derivation is straightforward, except that in order to determine the virtual work done, care must be taken to ensure that each distributed load on a member is multiplied by the average displacement of the member in the direction of the load.

Leaving aside the joint rotation mechanisms, which are only used in combinations, the work equations for the independent mechanisms, viewed as possible plastic collapse mechanisms, are found from the equilibrium equations to be:

$$\text{Beam AB } 37.5\lambda\theta = 80\theta; \qquad \lambda = 2.133$$

$$\text{Beam CD } 135\lambda\theta = 240\theta; \qquad \lambda = 1.778$$

$$\text{Beam-type AC } \quad 6\lambda\theta = 80\theta; \qquad \lambda = 13.3$$

$$\text{Beam-type CE } \quad 8\lambda\theta = 160\theta; \qquad \lambda = 20$$

$$\text{Beam-type BD } \quad 6\lambda\theta = 80\theta; \qquad \lambda = 13.3$$

$$\text{Beam-type DF } \quad 8\lambda\theta = 160\theta; \qquad \lambda = 20$$

$$\text{Sway ABCD } \quad 24\lambda\theta = 80\theta; \qquad \lambda = 3.333$$

$$\text{Sway CDEF } \quad 96\lambda\theta = 160\theta; \qquad \lambda = 1.667.$$

The lowest value of λ for these mechanisms is 1.667 for the sway mechanism CDEF.

The four beam-type mechanisms for the columns AC, CE, BD and DF have very high values of λ. It is concluded that they probably will not feature in any combination which will produce a lower value of λ than 1.667, and so they are disregarded in what follows.

The first combination considered is of the two mechanisms with the lowest values of λ, namely the beam CD and the sway of CDEF. A direct addition of these two mechanisms, as in Fig. 4.8(a), does not eliminate any hinges. However, at the joint C there are hinge rotations of magnitude θ in both CD and CE, so that the work absorbed at this joint is $60\theta + 40\theta = 100\theta$. By rotating this joint clockwise through an angle θ these hinges are both cancelled, and replaced by a hinge rotation $+\theta$ in AC, as shown in Fig. 4.8(b). This reduces the work absorbed at this joint to 20θ, a reduction of 80θ. A rotation of the joint D does not achieve any reduction in the work absorbed. The work equation for this combination is therefore obtained as follows:

$$\text{Sway CDEF } \quad 96\lambda\theta = 160\theta; \qquad \lambda = 1.667$$

$$\text{Beam CD } \quad \underline{135\lambda\theta = 240\theta}; \qquad \lambda = 1.778$$

$$\text{Combination Fig. 4.8(b) } \quad 231\lambda\theta = 400\theta - 80\theta = 320\theta$$

$$\lambda = 1.385.$$

If the sway mechanism of ABCD, Fig. 4.7(c), is added to the combination just established, there is a cancellation of the hinge at section 8, joint C. The resulting mechanism is shown in Fig. 4.8(c). The work absorbed at the hinge was 20θ in both mechanisms, so that the reduction in the work absorbed is 40θ. Again

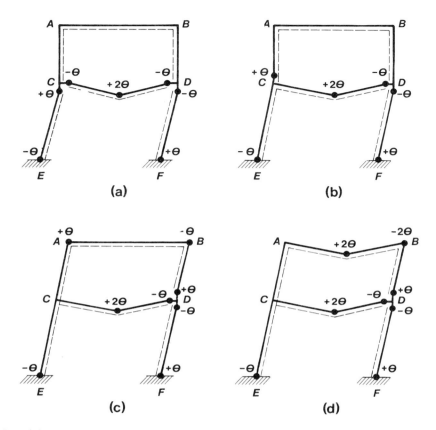

Fig. 4.8 *Two-storey rectangular frame: combinations of mechanisms*

there is no advantage in rotation of the joint D; a clockwise rotation θ leaves the total work absorbed at this joint unchanged. This implies that the mechanism of Fig. 4.8(c) is overcomplete, as is otherwise evident from the fact that it involves 8 hinges, the number of redundancies r being 6. The work equation is obtained as follows:

Combination Fig. 4.8(b) $231\lambda\theta = 320\theta$; $\lambda = 1.385$

Sway ABCD $\underline{24\lambda\theta = 80\theta}$; $\lambda = 3.333$

Combination Fig. 4.8(c) $255\lambda\theta = 400\theta - 40\theta = 360\theta$

$$\lambda = 1.412.$$

The beam mechanism AB is now added to this combination. The resulting mechanism is shown in Fig. 4.8(d), and it will be seen that there is a hinge cancellation at A. The work absorbed at A was 20θ in both mechanisms, so that there is a reduction in the work absorbed of 40θ. The work equation is thus:

Combination Fig. 4.8(c) $255\lambda\theta = 360\theta;$ $\lambda = 1.412$

Beam AB $\underline{37.5\lambda\theta = 80\theta;}$ $\lambda = 2.133$

Combination Fig. 4.8(d) $292.5\lambda\theta = 440\theta - 40\theta = 400\theta$

$$\lambda = 1.368.$$

This is the lowest value of λ which has been found. All possible combinations have not been explored; in particular those involving the beam-type mechanisms for the columns, and also the combination of beam AB with the sway of ABCD. However, the above combinations appeared to be those most likely to lead to low values of λ, and so it is concluded that the mechanism of Fig. 4.8(d) is the actual collapse mechanism, subject to any adjustments of the positions of the plastic hinges which at this stage are located at the mid-points of the beams AB and CD.

Before these adjustments are made, this conclusion is checked by a statical analysis. Details need not be given; the plastic moments involved in the mechanism of Fig. 4.8(d) are substituted in the equilibrium equations, and it is found that $\lambda = 1.368$, confirming the combining mechanisms calculation. The bending moment distribution is:

Section	1	2	3	4	5	6	7	8
M	$+8.72$	$+20$	-20	-4.62	$+60$	-60	$+16.41$	$+15.90$

Section	9	10	11	12	13	14	15	16
M	$+11.28$	-8.89	-40	-4.10	$+20$	-40	-5.47	$+40.$

The bending moments above are safe, since the plastic moment is not exceeded at any of the cross sections considered. This confirms that the mechanism of Fig. 4.8(d) is the actual collapse mechanism, subject to adjusting the positions of the plastic hinges in AB and CD. It is also necessary to check that the plastic moment is not exceeded anywhere within the spans of the other four members.

The maximum bending moment $M^{\max}$ in each member may be calculated using the results established in Section 3.4, namely

$$y_0 = (M_R - M_L)/W. \qquad (3.16)$$

$$M^{\max} = M_C + Wy_0^2/2L. \qquad (3.17)$$

Here L is the length of the member, W is the normal load and M_L, M_C and M_R are the bending moments at the left-hand end, the centre and the right-hand end, respectively. $M^{\max}$ occurs at a distance y_0 to the right of the centre of the member. The sign conventions were specified in Fig. 3.2. The results are given in Table 4.2.

Table 4.2 *Maximum bending moments in members*

Member	M_L	M_C	M_R	W	L	y_0	M^{max}	M_p
AB	+8.72	+20	−20	34.2	6	−0.84	+22.01	20
CD	−4.62	+60	−60	123.1	6	−0.45	+62.08	60
AC	+15.90	+16.41	+ 8.72	10.9	3	−0.66	+17.20	20
BD	−20	−4.10	+20	−10.9	3	−	−	20
CE	−40	−8.89	+11.28	10.9	4	−	−	40
DF	−40	−5.47	+40	−10.9	4	−	−	40

For BD, CE and DF the calculated values of y_0 are greater in magnitude than the semi-lengths of the members, so that there is no mathematical maximum bending moment in these members. For AC, the maximum bending moment is less than the plastic moment. It is only in AB and CD that maximum moments occur which are greater than the plastic moment, and the biggest discrepancy occurs in AB, where the maximum moment is 10 per cent greater than the plastic moment. Following the argument of Section 3.4, this establishes the following bounds on λ_c:

$$\left(\frac{20}{22.01}\right) 1.368 < \lambda_c < 1.368$$

$$1.243 < \lambda_c < 1.368.$$

To improve this result, the plastic hinges in AB and CD are moved to the positions of maximum bending moment given in Table 4.2, and a new value of λ is calculated by a kinematical analysis. Details will not be given; the result obtained is

$$\lambda = 1.342.$$

This value of λ may be taken to be λ_c for all practical purposes. The required values of the plastic moments for a load factor of 1.5 are then obtained by multiplying the trial values by $1.5/1.342$.

To obtain the exact value of λ_c, a work equation could be written down for the mechanism of Fig. 4.8(d) with the hinges in the beams AB and CD located at variable positions, and the corresponding value of λ minimized by differentiation. This analysis gives the value 1.342, agreeing with the above value to four significant figures. The values of y_0 are found to be −0.78 m for AB and −0.45 m for CD.

A general analysis of the problem of determining the correct positions of the plastic hinges within the members of multistorey, multibay rectangular frames, together with the corresponding load factor, has been given by Horne (1954a). However, the above procedure will be found to give results of sufficient accuracy; a summary of the steps involved is as follows:

(a) Determine the 'correct' collapse mechanism, assuming that any plastic hinges within those members which carry uniformly distributed loads occur at mid-span.

(b) Perform a statical check.

(c) Determine the positions and magnitudes of the maximum bending moments in the members which carry uniformly distributed loads.

(d) For the members which carry uniformly distributed loads in which plastic hinges occurred at mid-span in the 'correct' mechanism, move those hinges to the positions of maximum moment found under (c), and analyse the resulting mechanism kinematically. The corresponding load factor may then be assumed to be λ_c.

4.3.5 Lean-to frame

The final example of the combining mechanisms technique is the frame shown in Fig. 4.9(a). This raises a new issue which will become apparent when the combination of independent mechanisms is considered. The load factor required is 1.5; the members all have the same plastic moment and a trial value of 20 kN m is assumed. The frame is rigidly built-in at section 4, but is pinned to a rigid base at the other foot.

For this frame $n = 4$ and $r = 2$, so that there are $(n - r) = 2$ independent mechanisms. These are shown in Fig. 4.9(b) and (c). In Fig. 4.9(b) the sway motion is specified by a rotation θ about I, the instantaneous centre of rotation for beam 13, whereas in the beam mechanism of Fig. 4.9(c) the displacements and hinge rotations are given in terms of the rotation $-\phi$ of the hinge at section 1. The work equations for these two mechanisms are:

$$\text{Sway Fig. 4.9(b)} \quad 120\lambda\theta = 160\theta; \qquad \lambda = 1.333$$

$$\text{Beam Fig. 4.9(c)} \quad 40\lambda\phi = 80\phi; \qquad \lambda = 2.$$

The only combination to consider is an addition of these two mechanisms with cancellation of the hinge at section 1. This cancellation is achieved if $\phi = 3\theta$. The beam equation then becomes

$$120\lambda\theta = 240\theta,$$

and the work absorbed at the plastic hinge at section 1 is 60θ for both this mechanism and the sway mechanism. The work equation for the combined mechanism, which is shown in Fig. 4.9(d), is therefore derived as follows:

$$\text{Sway Fig. 4.9(b)} \quad 120\lambda\theta = 160\theta; \qquad\qquad\qquad \lambda = 1.333$$

$$\text{Beam Fig. 4.9(c)} \quad \underline{120\lambda\theta = 240\theta(\phi = 3\theta);} \qquad \lambda = 2$$

$$\text{Combination Fig. 4.9(d)} \quad 240\lambda\theta = 400\theta - 120\theta = 280\theta$$

$$\lambda = 1.167.$$

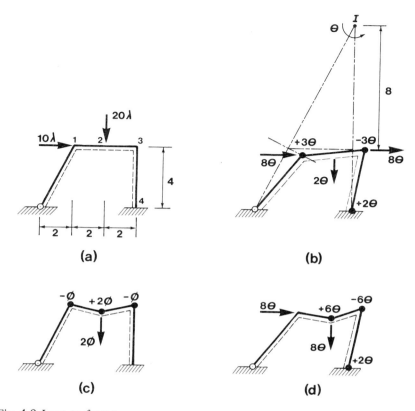

Fig. 4.9 *Lean-to frame*
    (a) Dimensions and loading. Units: m, kN
    (b) Sway mechanism
    (c) Beam mechanism
    (d) Combined mechanism

This is the lowest of the three values of λ, and so it is concluded that the mechanism of Fig. 4.9(d) is the actual collapse mechanism. A statical check is easily made in the usual way, and will not be given. The required value of the plastic moment is

$$\left(\frac{1.5}{1.167}\right) 20 = 25.7 \text{ kN m}.$$

The chief point to note about this calculation is the need to arrange for the hinge rotations at section 1 to be the same in magnitude, so that full cancellation is achieved. It should also be appreciated that the quite complicated kinematics of the combined mechanism of Fig. 4.9(d) did not have to be worked out from first principles; it was only necessary to add the displacements and hinge rotations of the two combining mechanisms.

4.4 Other methods for determining the collapse load factor

The combining mechanisms technique represents an upper bound, or kinematic approach to the determination of λ_c for a given frame and loading. Lower bound methods were developed simultaneously by Horne (1954b) and English (1954); Horne's method of plastic moment distribution, as its name implies, has some features in common with the well-known elastic moment distribution procedure for plane frames.

The lower bound approach is readily expressed in a form suitable for solution by digital computing techniques. Consider for example the frame of Fig. 4.5(a). The equations of equilibrium for this frame were shown in Section 4.3.2 to be

$$100\lambda = -M_1 + M_2 - M_{10} + M_5 + M_9 - M_8. \tag{4.18}$$

$$60\lambda = -M_2 + 2M_3 - M_4. \tag{4.19}$$

$$72\lambda = -M_6 + 2M_7 - M_8. \tag{4.20}$$

$$0 = -M_4 - M_5 + M_6. \tag{4.21}$$

The ten bending moments cannot exceed the plastic moment 30 kN m in magnitude, so that

$$-30 \leqslant M_i \leqslant 30 \qquad (i = 1, 2, \ldots, 10). \tag{4.32}$$

The collapse load factor λ_c is then the highest value of λ which can be found, subject to the four constraints imposed by the requirements of equilibrium, Equations (4.18)–(4.21), and the twenty constraints imposed by the requirements of yield, inequalities (4.32).

This formulation of the problem was used by Neal and Symonds (1950–1951) in developing a method of analysis based on the solution of systems of linear inequalities by a procedure due to Dines (1918–19). However, it represents a standard problem in linear programming, and was first identified as such by Charnes and Greenberg (1951). Further discussions have been given by various authors, including Livesley (1956), Prager (1957) and Heyman (1959).

The duality principle of linear programming may be used to formulate procedures which are upper bound or kinematic in character. This was pointed out by Dorn and Greenberg (1957), Charnes, Lemke and Zienkiewicz (1959) and also Hoskin (1960). A procedure for developing simultaneous upper and lower bounds was developed by Munro (1965). A full discussion of the relationships between the static and kinematic theorems and primal-dual linear programs has been given by Munro and Smith (1972).

References

Baker, J.F. (1949), 'The design of steel frames', *Struct. Engr*, **27**, 397.

Baker, J.F. and Eickhoff, K.G. (1955), 'The behaviour of saw-tooth portal frames', prelim. vol., Conf. Correlation between Calculated and Observed Stresses and Displacements in Structures, Inst. Civil. Engrs, 107.

Baker, J.F. and Eickhoff, K.G. (1956), 'A test on a pitched roof portal frame', prelim. publ., 5th Congr. Int. Assoc. Bridge Struct. Eng., Lisbon, 1956.

Charnes, A. and Greenberg, H.J. (1951), 'Plastic collapse and linear programming', *Bull. Am. Math. Soc.*, **57**, 480.

Charnes, A., Lemke, C.E. and Zienkiewicz, O.C. (1959), 'Virtual work, linear programming and plastic limit analysis', *Proc. R. Soc.*, A, **251**, 110.

Dines, L.L. (1918–19), 'Systems of linear inequalities', *Ann. Math., Princeton* (series 2), **20**, 191.

Dorn, W.S. and Greenberg, H.J. (1957), 'Linear programming and plastic limit analysis of structures', *Q. Appl. Math.*, **15**, 155.

Driscoll, G.C. and Beedle, L.S. (1957), 'The plastic behaviour of structural members and frames', *Weld. J., Easton, Pa.*, **36**, 275-s.

English, J.M. (1954), 'Design of frames by relaxation of yield hinges', *Trans. Am. Soc. Civ. Engrs*, **119**, 1143.

Harrison, H.B. (1960), 'The preparation of charts for the plastic design of mild steel portal frames', *Civ. Eng. Trans., Inst. Engrs Austr.*, March.

Hendry, A.W. (1955), 'Plastic analysis and design of mild steel Vierendeel girders', *Struct. Engr*, **33**.213.

Heyman, J. (1957), *'Plastic design of portal frames'*, Cambridge University Press.

Heyman, J. (1959), 'Automatic analysis of steel framed structures under fixed and varying loads', *Proc. Inst. Civil Engrs*, **12**, 39.

Horne, M.R. (1954a), 'Collapse load factor of a rigid frame structure', *Engineering*, **177**, 210.

Horne, M.R. (1954b), 'A moment distribution method for the analysis and design of structures by the plastic theory', *Proc. Inst. Civil Engrs*, **3**, (part 3), 51.

Hoskin, B.C. (1960), 'Limit analysis, limit design and linear programming', Aeronautical Research Laboratories, Melbourne, Report ARL/SM. 274.

Livesley, R.K. (1956), 'The automatic design of structural frames', *Q. J. Mech. Appl. Math.*, **9**, 257.

Munro, J. (1965), 'The elastic and limit analysis of planar skeletal structures', *Civ. Eng. Publ. Wks Rev.*, **60**, May.

Munro, J. and Smith, D.L. (1972), 'Linear programming duality in plastic analysis and synthesis', *Proc. Int. Symp. Computer-aided Structural Design*, vol. 1, Warwick Univ.

Neal, B.G. and Symonds, P.S. (1950–51), 'The calculation of collapse loads for framed structures', *J. Inst. Civil Engrs*, **35**, 21.

Neal, B.G. and Symonds, P.S. (1952a), 'The rapid calculation of the plastic collapse load for a framed structure', *Proc. Inst. Civil Engrs*, **1**, (part 3), 58.

Neal, B.G. and Symonds, P.S. (1952b), 'The calculation of plastic collapse loads for plane frames', prelim. publ., 4th Congr. Int. Assoc. Bridge Struct. Eng., Cambridge, 1952, 75. (Reprinted in *Engineer*, **194**, 315, 363.)

Prager, W. (1957), 'Linear programming and structural design: I. Limit analysis; II. Limit design', Papers P-1122, 1123. Rand Corporation.

Examples

1. In the fixed-base, pitched-roof portal frame of Fig. 4.1(a), the uniformly distributed vertical loads on the rafters, shown as 18λ kN and 26λ kN, are reduced to 9λ kN and 17λ kN, respectively, the other loads remaining unchanged. If all the members of the frame have the same plastic moment 30 kN m, find the collapse load factor.

2. In the fixed-base, pitched-roof portal frame of Fig. 4.1(a), the uniformly distributed vertical loads on the rafters, shown as 18λ kN and 26λ kN, are increased to 28λ kN each. All other loads are zero, and all members of the frame have the same plastic moment 60 kN m. Find the collapse load factor. If both the feet are pinned rather than fixed, what is the new collapse load factor?

3. A fixed-base, pitched-roof portal frame ABCDE consists of two columns AB and ED, each of length 4.5 m and with the feet A and E 12 m apart, together with two rafters BC and DC of equal length and inclined at $15°$ to the horizontal. All members of the frame have the same plastic moment 60 kN m.

If each rafter carries a uniformly distributed vertical load 35λ kN, find the collapse load factor, and show that it is unchanged if in addition there is a uniformly distributed horizontal load 15λ kN acting on the column AB in the direction AE.

4. A pinned-base, saw-tooth frame ABCDE has two columns AB and ED, each of length 3.6 m and with the feet A and E 7.8 m apart. The rafters BC and DC are of length 7.2 m and 3 m, respectively.

The rafter BC carries a uniformly distributed vertical load 40λ kN. If all members of the frame have the same plastic moment 25 kN m, find the collapse load factor.

5. A lean-to, fixed-base frame ABCD consists of two columns AB and DC, whose lengths are 3 m and 3.9 m, respectively, the feet A and D being 4.8 m apart, together with a rafter BC. All members of the frame have the same plastic moment 25 kN m.

If the rafter BC carries a uniformly distributed vertical load 50λ kN, find the collapse load factor.

If in addition there is a uniformly distributed horizontal wind load 15λ kN on the column AB in the direction AD, and also a uniformly distributed wind suction 2.5λ kN on the rafter BC normal to this member, find the new value of the collapse load factor.

6. A pinned-base, pitched-roof portal frame ABCDE has columns AB and ED each of height 3 m, the feet A and E being 12 m apart. The rafters BC and DC

are of equal length and inclined at $22\frac{1}{2}°$ to the horizontal. The knees B and D are connected by a tie-rod which prevents any relative horizontal movement but cannot sustain any appreciable bending moment. All members of the frame have the same plastic moment M_p. Each rafter carries a uniformly distributed vertical load 50λ kN. Find the value of M_p which provides a load factor of 1.5 against plastic collapse. What is the tension in the tie-rod at collapse?

7. In a two-bay, fixed-base base rectangular frame ABCDEF the three columns AB, FC and ED are each of length 4 m and plastic moment 30 kN m, while the two beams BC and CD are each of length 5 m and plastic moment 60 kN m. A horizontal load H acts at B in the direction BC, and in addition there are concentrated vertical loads P and Q at the centres of the beams BC and CD respectively.

Find the collapse load factors for the load combinations:

(a) $H = 25\lambda$ kN, $P = 40\lambda$ kN, $Q = 40\lambda$ kN.
(b) $H = 17.5\lambda$ kN, $P = 42\lambda$ kN, $Q = 56\lambda$ kN.

Estimate the change in the collapse load factor in case (a) if the loads P and Q are changed to 80λ kN uniformly distributed over each beam.

8. In a two-storey, single-bay, fixed-base rectangular frame ABCDEF the continuous columns ABC and FED are each of total length 8 m, and $AB = BC = DE = EF = 4$ m. The feet A and F are 7.2 m apart, so that the upper and lower beams CD and BE each span 7.2 m.

There are concentrated vertical loads V_1 and V_2 at the centres of the beams CD and BE, respectively, and concentrated horizontal loads H_1 and H_2 at D and E, respectively, acting in the directions CD and BE. All members of the frame have the same plastic moment 40 kN m.

Find the collapse load factors for the load combinations:

(a) $V_1 = 20\lambda$ kN, $V_2 = 20\lambda$ kN, $H_1 = 10\lambda$ kN, $H_2 = 10\lambda$ kN.
(b) $V_1 = 30\lambda$ kN, $V_2 = 30\lambda$ kN, $H_1 = 0$, $H_2 = 15\lambda$ kN.

9. In a three-storey, single-bay, fixed-base rectangular frame ABCDEFGH, each storey is of height 3 m and the span of each beam is also 3 m. The plastic moments of the members are as follows:

Upper storey columns CD and FE: 30 kN m

Middle storey columns BC and GF: 60 kN m

Lower storey columns AB and HG: 90 kN m

Beam DE: 30 kN m

Beams CF and BG: 60 kN m

Concentrated horizontal loads 10λ kN, 20λ kN and 30λ kN are applied at E, F and G respectively, all these loads acting in the same direction. Find the collapse load factor.

10. In a multibay, fixed-base, pitched-roof portal frame all the bays are of identical shape, with column height h, span l and rafter slope θ. All members have plastic moment M_p. Each rafter carries a uniformly distributed vertical load λW.

Show that collapse is confined to the outermost bay at each end of the frame, and that the collapse load factor is $4M_p(2h + l \tan \theta)/Wlh$, neglecting the small correction due to the fact that plastic hinges do not form at the apices of the outermost frames but instead at a small distance away from these apices.

11. A semi-circular arch of radius R is of uniform cross section, with plastic moment M_p, and is pinned at both feet to rigid abutments. It carries a central vertical concentrated load W. Find the horizontal abutment thrust and the value of W at collapse, neglecting the effect of axial thrust on the plastic moment.

12. In a two-bay, fixed-base rectangular frame, each bay has a span 8 m. One bay ABCDE is of height 8 m while the other bay EDFG is of height 4 m. The column EDC is common to both bays and E is its base: ED = DC = 4 m. The taller bay has a column AB of length 8 m, and the other bay has a column GF of length 4 m, the bases being A and G, respectively. All members of the frame have the same plastic moment 48 kN m.

The beams BC and DF, each of span 8 m, carry central concentrated vertical loads 9λ kN and 27λ kN, respectively. Horizontal concentrated loads 18λ kN and 9λ kN are applied at C and F, respectively, both acting in the same direction BC. Find the collapse load factor.

5 Estimates of Deflections

5.1 Introduction

The plastic methods described in Chapters 3 and 4 are concerned solely with determining the strength of frames. However, it is possible that excessive deflections might occur in a frame before the plastic collapse load was attained, rendering the structure unserviceable. The design would then need to be based on a serviceability criterion and load factor, rather than on the collapse load factor. There is therefore a need for methods which enable the deflections of a frame to be calculated at the point of collapse, and the purpose of this chapter is to describe such methods.

A further reason for discussing this question is that the plastic theory assumes that the deflections developed in a frame prior to collapse have a negligible effect on its geometry, so that the equations of equilibrium are sensibly those for the undistorted frame. It may therefore be advisable in some cases to estimate the deflections at the point of collapse to see whether they are sufficiently large to invalidate the assumption of unaltered geometry.

In much of this chapter proportional loading will be assumed. However, in practice a structure may experience variable and repeated loading, as in the case of a building frame which may be subjected to several different severe combinations of wind and snow loads during its lifetime. It will be seen in Chapter 8 that this type of loading can cause a progressive increase in the deflections, even if the peak values of the loads are always appreciably less than those which would cause plastic collapse. This should be borne in mind when assessing the value of deflection estimates obtained under the assumption of proportional loading.

It is commonly assumed in the elastic analysis of beams and plane frames that deflections due to shear and axial forces are negligible by comparison with those due to bending. If this is accepted for partially plastic behaviour, the relationship between load and deflection can in principle be determined once the bending moment-curvature relation is specified. Some examples of this process are given in Section 5.2 for simply supported beams of rectangular cross section, assuming ideal plasticity. The calculations become extremely cumbersome for structures of any degree of complexity, and approximations must be made to produce a workable method. These are discussed in Section 5.3, and a method for estimating the deflections of a frame at the point of collapse is then given in Section 5.4.

5.2 Load-deflection relations for simply supported beams

5.2.1 *Rectangular cross section: ideal plastic material*

Consider a beam of rectangular cross section, breadth B and depth D, which is simply supported over a span l. The beam is initially assumed to carry a central concentrated load W, and the bending moment diagram is then as shown in Fig. 5.1.

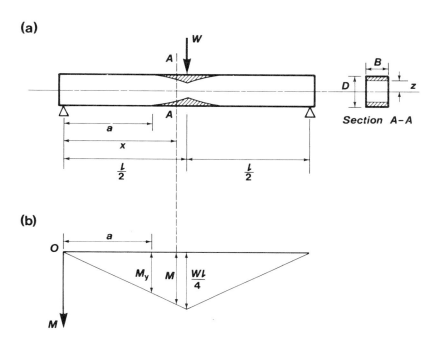

Fig. 5.1 *Simply supported rectangular section beam with central concentrated load*

    (a) Beam and loading: plastic zones shaded

    (b) Bending moment diagram

It will be assumed that the beam is of ideal plastic material, initially free from stress, and that the load has been steadily increased until plasticity has developed in the regions shown. The bending moment-curvature relation for a beam of this cross section and material was developed in Section 1.3.1, and can be summarized as follows:

$$\text{elastic} \quad \frac{M}{M_y} = \frac{\kappa}{\kappa_y}, \quad 0 \leqslant x \leqslant a \qquad (5.1)$$

$$\text{yielded} \quad \frac{M}{M_y} = 1.5 - 0.5\left(\frac{\kappa_y}{\kappa}\right)^2, \quad a \leqslant x \leqslant \frac{l}{2} \tag{5.2}$$

$$\frac{2z}{D} = \frac{\kappa_y}{\kappa}. \tag{5.3}$$

Equation (5.2) was developed for the case of pure bending; it will now be assumed that it is applicable when shear forces are present.

From the bending moment diagram it is seen that

$$M_y = \tfrac{1}{2}Wa. \tag{5.4}$$

The central bending moment is $Wl/4$. Thus, if W_0 denotes the value of W at which yield first occurs at the centre of the beam,

$$M_y = \tfrac{1}{4}W_0 l, \tag{5.5}$$

so that

$$a = \frac{l}{2}\left(\frac{W_0}{W}\right). \tag{5.6}$$

In addition,

$$M = \tfrac{1}{2}Wx, \quad \text{for} \quad 0 \leqslant x \leqslant \frac{l}{2}, \tag{5.7}$$

and combining this result with Equation (5.4),

$$\frac{M}{M_y} = \frac{x}{a}. \tag{5.8}$$

Using this equation in combination with Equations (5.1) and (5.2), the curvature can now be expressed as a function of x as follows:

$$\kappa = \kappa_y\left(\frac{x}{a}\right), \qquad 0 \leqslant x \leqslant a$$

$$\kappa = \kappa_y\left(3 - 2\frac{x}{a}\right)^{-1/2}, \quad a \leqslant x \leqslant \frac{l}{2}. \tag{5.9}$$

The central deflection δ is found by the unit load method (Section 2.5.5) to be given by

$$\delta = \int_0^{\frac{l}{2}} x\kappa\,dx \tag{5.10}$$

$$= \int_0^a \frac{\kappa_y}{a}x^2\,dx + \int_a^{\frac{l}{2}} \kappa_y x\left(3 - 2\frac{x}{a}\right)^{-1/2}\,dx.$$

Evaluating these integrals, and eliminating a by using Equation (5.6),

$$\delta = \frac{l^2 \kappa_y}{12} \left(\frac{W_0}{W}\right)^2 \left[5 - \left(3 + \frac{W}{W_0}\right)\left(3 - 2\frac{W}{W_0}\right)^{1/2}\right].$$

The deflection δ_0 when yield first occurs at the centre of the beam is obtained by putting $W = W_0$, giving

$$\delta_0 = \frac{l^2 \kappa_y}{12}.$$

It follows that

$$\frac{\delta}{\delta_0} = \left(\frac{W_0}{W}\right)^2 \left[5 - \left(3 + \frac{W}{W_0}\right)\left(3 - 2\frac{W}{W_0}\right)^{1/2}\right], \tag{5.11}$$

a result first obtained by Fritsche (1930).

Failure by plastic collapse will occur when the central bending moment $Wl/4$ reaches the plastic moment $M_p = 1.5M_y$. The collapse load W_c is therefore given by

$$M_p = \tfrac{1}{4}W_c l,$$

and comparing this with Equation (5.5)

$$\frac{W_c}{W_0} = \frac{M_p}{M_y} = 1.5.$$

The deflection δ_c at the point of collapse is then found from Equation (5.11), with $W = 1.5W_0$, to be $2.22\delta_0$.

The load-deflection relation is plotted in Fig. 5.2, curve (i). Once the collapse load $W_c = 1.5W_0$ has been attained, the deflection can grow indefinitely due to the rotation of the central plastic hinge. However, the deflection developed before this load is reached is finite and of the same order as the elastic deflection at the point of first yield.

The shape of the boundaries of the yielded zones can be found from Equations (5.3) and (5.9), which when combined give

$$\frac{2z}{D} = \left(3 - 2\frac{x}{a}\right)^{1/2}$$

Using Equation (5.6) to eliminate a,

$$\left(\frac{2z}{D}\right)^2 = 3 - \frac{4x}{l}\left(\frac{W}{W_0}\right),$$

so that in this case the boundaries are parabolic. Fig. 5.3(a) shows the shape of these zones when $W = W_c$.

The analysis can readily be extended to cover the case of symmetrical two-point loading, as in Fig. 5.3(b). Details will not be given, but attention will be drawn to the salient features. The central portion of the beam is subjected to a

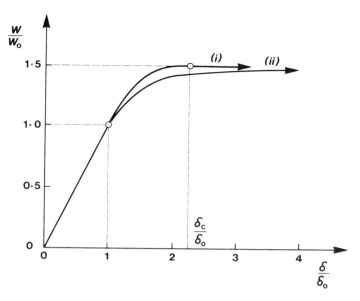

Fig. 5.2 *Load-deflection relations for simply supported beam of rectangular cross section*

(i) Central concentrated load
(ii) Symmetrical two-point load

constant bending moment and therefore bends into an arc of a circle. The shape of the yielded zones is shown in the figure for the case in which the distance between the loads is one-third of the span, and the central bending moment is the plastic moment M_p. The load-deflection relation is shown in Fig. 5.2, curve (ii). It will be seen that the collapse load is only reached when the deflection becomes infinite, by contrast with the previous case of a central concentrated load. This is because at collapse the plastic moment is attained over a finite length of the beam rather than at a single cross section. The infinite curvature associated with the plastic moment therefore causes an infinite deflection.

If the beam carries a uniformly distributed load, the distribution of bending moment is parabolic, and is shown in Fig. 5.3(c) at the collapse load. The yielded zones in this case have boundaries which are linear. The load-deflection relation has the same feature as that for the two-point loading, the deflection becoming infinite as the collapse load is attained. This is due to the form of the bending moment diagram. At the centre of the beam the shear force and thus the rate of change of bending moment with distance along the beam is zero. This approximates more nearly to the condition of pure bending over a finite length of the beam, as in Fig. 5.3(b), than the linear rate of change of bending moment with distance along the beam depicted in Fig. 5.3(a).

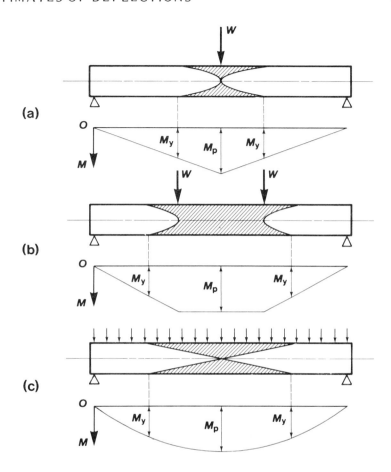

Fig. 5.3 *Shape of plastic zones for simply supported beams of rectangular cross section at collapse*

    (a) Central concentrated load
    (b) Symmetrical two-point load
    (c) Uniformly distributed load

5.2.2 *Other cross sections and material properties*

The analyses just described assumed the relationship between bending moment M and curvature κ which was derived in Section 1.3.1 for a rectangular cross section beam of ideal plastic material. This derivation assumed a linear variation of longitudinal strain ϵ across the section. If this assumption is retained, the (M, κ) relation can be determined for any assumed relationship between longitudinal stress σ and strain ϵ, and for any cross section. However, it will be recalled from Section 1.2 that for mild steel the yielding process is discontinuous, so that only the average strain over a finite length of the beam can be presumed to vary linearly across the section.

As pointed out in Section 1.2, the upper yield phenomenon occurs in specimens of annealed mild steel, but is destroyed by cold-working, and is not normally present in rolled steel sections. It is usually taken into account only in careful experimental work aimed at verifying fundamental theory. Investigations of this type were carried out on beams of rectangular cross section by Cook (1937) and by Roderick and Phillipps (1949). Roderick and Heyman (1951) extended the theory to include the effect of strain-hardening. These investigations provided convincing evidence in support of the assumptions made.

Dwight (1953) considered non-linear (σ, ϵ) relations appropriate to aluminium alloys, and obtained good agreement with experiments on beams of rectangular cross section. In this work the beams were bent about a principal axis; difficulties arise when bending takes place about an axis other than the principal axis, and these were discussed by Barrett (1953).

Several investigations have been carried out with the purpose of correlating the observed load-deflection behaviour of beams of I-section with actual (σ, ϵ) relations obtained from material tests. These include the work of Roderick (1954), Roderick and Pratley (1954) and Sawyer (1961). The presence of residual stresses obviously influences the results, and their effect on the (M, κ) relation has been studied by several investigators, including Young and Dwight (1971).

5.3 Effects of strain-hardening and shape factor

The effect of strain-hardening on the load-deflection behaviour of beams of I-section was studied by Hrennikoff (1948). The assumed (σ, ϵ) relation was as shown in Fig. 5.4(a), in which the material properties characterizing the onset of strain-hardening were as follows:

$$\epsilon_s = 16.4\epsilon_0$$

$$E_s = E/48.$$

Hrennikoff assumed that the thickness of each flange was negligible as compared with the depth of the beam, so that each flange area could be regarded as concentrated at a constant distance from the neutral axis. The form of the (M, κ) relation then depends only on the ratio of the total flange area A_f to the web area A_w. Taking this ratio as unity, it can be shown that the shape factor ν is 1.125, so that $M_p = 1.125M_y$. The corresponding (M, κ) relation is shown in Fig. 5.4(b). It will be seen that strain-hardening commences when $\kappa = 16.4\kappa_y$, this being the curvature at which ϵ reaches the value ϵ_s in the outermost fibres.

This (M, κ) relation was used by Horne (1948) to analyse the pin-based rectangular frame of height l and span $2l$ shown in Fig. 5.5. The loading programme was first to bring the vertical load V up to the value $2.84M_p/l$, which just causes the yield moment to be attained at the most highly stressed cross section 3. V was then held constant while H was increased steadily. The relation between H and the corresponding deflection h is shown as curve (i) in Fig. 5.5.

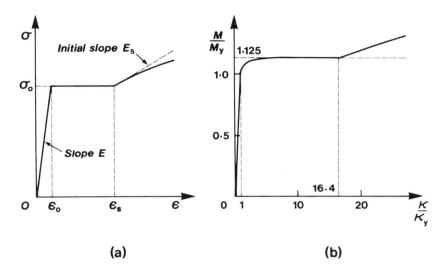

(a)   **(b)**

Fig. 5.4 *Bending moment-curvature relation for* I-*section with strain-hardening*
(a) Stress-strain relation
(b) Bending moment-curvature relation: $A_f = A_w$

The behaviour is elastic until $H = 0.71 M_p/l$, when the yield moment is attained at section 4. While the frame is responding elastically to H, the bending moment at section 3 does not change, and so remains at the yield value. As H increases further, yield spreads into the webs of the members at sections 3 and 4 and also along the members for some distance from these sections, and the slope of the load-deflection relation decreases steadily. Strain-hardening commences at section 4 when $H = 1.02 M_p/l$, and at section 3 when $H = 1.18 M_p/l$.

It is readily verified that, on the simple plastic theory, plastic collapse occurs with plastic hinges at sections 3 and 4 when $H = 1.16 M_p/l$, with V held at the value $2.84 M_p/l$. The indefinite increase of deflection under constant load predicted by the simple theory is seen from Fig. 5.5 to be prevented by strain-hardening. Nevertheless, the deflections do increase rapidly with small increases of the load above the predicted collapse load.

It is instructive to compare this load-deflection relation with two others which are shown in Fig. 5.5. Curve (ii) shows the effect of neglecting strain-hardening ($E_s = 0$). It differs little from curve (i) until $H = 1.16 M_p/l$, the plastic collapse load predicted by the simple plastic theory.

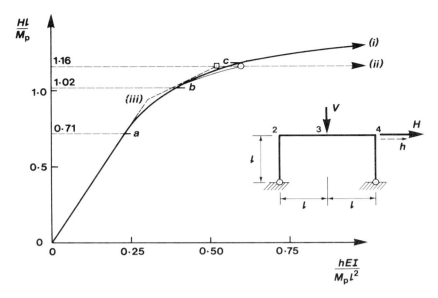

Fig. 5.5 *Load-deflection relations for pin-based rectangular frame*

    (i) Strain-hardening considered
      (a) Yield (b) Start of strain-hardening at 4
      (c) Start of strain-hardening at 3
    (ii) Strain-hardening ignored
    (iii) Ideal (M, κ) relation assumed

Curve (iii) also neglects strain-hardening, and in addition neglects the influence of the spread of plastic zones along the members. This is equivalent to assuming a shape factor of unity, so that each member is assumed to remain elastic until the plastic moment is reached, as in the step-by-step calculations of Chapter 2 (see Fig. 2.1).

At the plastic collapse load the calculated deflections are as follows:

    (a) $v = 1.125$ $E_s = 48$ $h = 0.56 M_p l^2 / EI.$
    (b) $v = 1.125$ $E_s = 0$ $h = 0.61 M_p l^2 / EI.$
    (c) $v = 1$ $E_s = 0$ $h = 0.53 M_p l^2 / EI.$

(a) may be considered to represent closely the way in which an actual structure would behave. The neglect of strain-hardening alone, as in (b), increases the deflection at the point of collapse by 9 per cent. The combined effects of neglecting strain-hardening and assuming a shape factor of unity, so that there is no spread of plastic zones along the members, as in (c), reduces this deflection by 5 per cent.

In general, the changes in the deflections at the point of collapse which can be attributed to strain-hardening and to the spread of plastic zones will be small.

Since these two effects tend to cancel one another, it is reasonable to neglect them. This forms the basis of the approximate method for estimating deflections at the point of collapse which is described in Section 5.4.

Several other studies of the problem of determining load-deflection relations for redundant structures have been made. Horne (1951) considered the effect of strain-hardening for built-in beams of both rectangular and I-section. Rawlings (1956) analysed rectangular frames whose members were of rectangular cross section, taking the upper yield stress into account. The observed behaviour of pinned and fixed-base rectangular frames of I-section was compared by Roderick (1960), using an analysis which took account of strain-hardening.

5.4 Estimates of deflections at point of collapse

5.4.1 *Assumptions*

The estimation of deflections at the point of collapse is greatly simplified if the effects of strain-hardening and the spread of plastic zones are neglected. This is unlikely to lead to serious error, as pointed out in Section 5.3. A further assumption which is made for proportional loading is that as the loads are increased to their collapse values, the rotation at a plastic hinge never ceases once it has formed. This implies that at any cross section,

$$-M_p < M < M_p, \qquad \theta = 0$$
$$M = -M_p, \qquad \theta < 0 \qquad\qquad (5.12)$$
$$M = M_p, \qquad \theta > 0.$$

This is not necessarily valid, even for proportional loading, as pointed out by Finzi (1957), and for more arbitrary loading programmes it can obviously be incorrect. Nevertheless, deflection estimates based on this assumption have been found to be of value, and the only alternative would be to trace the complete behaviour of a frame, for a given programme of loading to collapse, by the laborious step-by-step process described in Section 2.5.

Acceptance of this assumption implies that each plastic hinge involved in the collapse mechanism will form in turn and then continue to undergo rotation. Thus just before the collapse load is attained, all except one of these hinges will have formed and undergone rotation. At the point of collapse, the bending moment at the position of the last hinge to form reaches the plastic value, but before motion of the collapse mechanism ensues the rotation at this hinge will be zero. The identification of the last hinge to form is the key to the various methods which have been evolved.

When the effects of strain-hardening and the spread of plastic zones are neglected, the members of a frame at the point of collapse will be elastic everywhere except at the plastic hinges. The deflections and hinge rotations can therefore be calculated by adapting the techniques of elastic structural analysis to take into account the conditions (5.12). The first method to be proposed, by Symonds and Neal (1951, 1952), was based on the slope-deflection technique. However, a virtual work approach is advantageous, as first pointed out by Tanaka (1961) and Heyman (1961), and will be used here.

5.4.2 *Basic equations*

If a frame subjected to concentrated loads has n unknown bending moments and r redundancies, there are

$$(n - r) \text{ independent equations of equilibrium}$$

and r independent equations of compatibility.

As explained in Section 2.5.2, the equations of equilibrium can be found by the principle of virtual displacements, using the independent mechanisms.

The equations of compatibility can be found by the principle of virtual forces. The procedure was explained in Section 2.5.3, and is summarized here for convenience. The relevant form of the principle of virtual forces is

$$\int \frac{m^*M}{EI} \, ds + \sum m^* \phi = 0 \tag{5.13}$$

where

$$M = \text{actual bending moment}$$

$$M/EI = \text{actual curvature}$$

$$\phi = \text{actual hinge rotation}$$

$$m^* = \text{hypothetical residual moment.}$$

The integration covers all members of the frame, and the summation covers all positions where plastic hinge rotation has occurred.

The integral is evaluated by noting that for a typical uniform straight segment AB of length L, within which both m^* and M vary linearly,

$$\int_B^A \frac{m^*M}{EI} \, ds = \frac{L}{6EI} \left[m_A^*(2M_A + M_B) + m_B^*(2M_B + M_A) \right]. \tag{5.14}$$

A particular deflection δ may be found by the unit load form of the principle of virtual forces. A virtual unit load corresponding to δ is imagined to be applied to the frame, and any virtual bending moment distribution M^* satisfying the requirements of equilibrium with this load is obtained. As explained in Section 2.5.5, this gives the result

$$\delta = \int \frac{M^*M}{EI} \, ds + \sum M^* \phi, \tag{5.15}$$

where the symbols have the same meaning as for Equation (5.13). The integral can be evaluated using Equation (5.14), with M^* replacing m^*.

5.4.3 Fixed-ended beam with off-centre load

The first example to be considered is the fixed-ended beam of length $3l$ which is subjected to a load W at a distance $2l$ from one end, as shown in Fig. 5.6(a). The beam is of uniform section with flexural rigidity EI and plastic moment M_p. For this beam

$$n = 3$$

$$r = 2 = \text{number of independent equations of compatibility}$$

$$n - r = 1 = \text{number of independent equations of equilibrium.}$$

The equation of equilibrium is obtained from the mechanism of Fig. 5.6(b) by the method of virtual displacements, and is

$$-M_1 + 3M_2 - 2M_3 = 2Wl. \tag{5.16}$$

This mechanism, with plastic hinges replacing the hinges shown, is the collapse mechanism, and at collapse

$$M_1 = -M_p, \quad M_2 = M_p, \quad M_3 = -M_p$$

$$W_c = 3\frac{M_p}{l}.$$

During collapse the shapes of the two segments 12 and 23 remain unchanged, the increases of deflection being due solely to the rotations at the three plastic hinges. Fig. 5.6(c) shows the corresponding deflected form of the beam, the hinge rotations being the total rotations which have occurred both before and during collapse. This situation is now analysed, and the two compatibility equations are obtained by using two hypothetical residual moment distributions m^* in Equation (5.13).

From Equation (5.16) it follows that residual moments must satisfy

$$-m_1 + 3m_2 - 2m_3 = 0. \tag{5.17}$$

This equation expresses the fact that any distribution of residual moment must vary linearly across the beam, as in Fig. 5.6(d). This is otherwise obvious because the shear force is constant over the whole span when $W = 0$.

Since $r = 2$, any residual moment distribution can be formed as a linear combination of two independent distributions. Two possible independent distributions (i) and (ii) are entered in the first two rows of Table 5.1, and are shown in Fig. 5.6(e). Any other distribution can be formed as a linear combination of (i) and (ii).

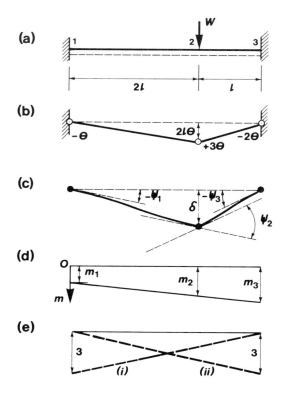

Fig. 5.6 *Fixed-ended beam with off-centre load*

    (a) Dimensions and loading
    (b) Virtual mechanism
    (c) Deformations during collapse
    (d) Residual moment distribution
    (e) Independent residual moment distributions

The deflection δ corresponding to W is determined from Equation (5.15), using any distribution of bending moment M^* in equilibrium with a unit load at section 2. This can be found from Equation (5.16) with $W = 1$; one possibility, shown as distribution (iii) in Table 5.1, is $M_2^* = M_3^* = 0$, so that $M_1^* = -2l$.

Table 5.1 is completed by the inclusion of the actual displacement system during collapse. Using the residual moment distribution (i) in Equation (5.13) and evaluating the integral with the aid of Equation (5.14),

$$\frac{2l}{6EI}[3(-2M_{\mathrm{p}} + M_{\mathrm{p}}) + (2M_{\mathrm{p}} - M_{\mathrm{p}})] + \frac{l}{6EI}[2M_{\mathrm{p}} - M_{\mathrm{p}}]$$

$$+ 3\psi_1 + \psi_2 = 0$$

Table 5.1 *Virtual force and actual displacement systems for beam of Fig. 5.6*

Section		1	2	3
Virtual force systems				
m^*	(i)	3	1	0
	(ii)	0	2	3
M^*	(iii)	$-2l$	0	0
Actual displacement system				
$EI\kappa = M$		$-M_p$	M_p	$-M_p$
ϕ		ψ_1	ψ_2	ψ_3

$$-\frac{M_p l}{2EI} + 3\psi_1 + \psi_2 = 0. \tag{5.18}$$

Proceeding similarly with distribution (ii), it is found that

$$\frac{M_p l}{2EI} + 2\psi_2 + 3\psi_3 = 0. \tag{5.19}$$

These are the two equations of compatibility. An expression for δ is found using distribution (iii), in conjunction with Equation (5.15); this gives

$$\delta = \frac{2M_p l^2}{3EI} - 2l\psi_1. \tag{5.20}$$

Equations (5.18)–(5.20) are insufficient in themselves to determine the four unknowns ψ_1, ψ_2, ψ_3 and δ. A way must be found of identifying the last hinge to form at the point of collapse. One method is to solve these three equations to give

$$\psi_1 = -\frac{\delta}{2l} + \frac{M_p l}{3EI}. \tag{5.21}$$

$$\psi_2 = \frac{3\delta}{2l} - \frac{M_p l}{2EI}. \tag{5.22}$$

$$\psi_3 = -\frac{\delta}{l} + \frac{M_p l}{6EI}. \tag{5.23}$$

These three equations hold true at any stage during collapse. It will be seen that the first term in each equation represents the mechanism motion during collapse, since if changes are denoted by the prefix Δ,

$$\Delta\psi_1 = -\frac{\Delta\delta}{2l}, \quad \Delta\psi_2 = \frac{3\Delta\delta}{2l}, \quad \Delta\psi_3 = -\frac{\Delta\delta}{l}.$$

The last hinge to form can now be identified by noting that each hinge rotation must have the same sign as the corresponding plastic moment given in Table 5.1. It follows that

$$\psi_1 \leqslant 0 \qquad \delta \geqslant \frac{2M_p l^2}{3EI}$$

$$\psi_2 \geqslant 0 \qquad \delta \geqslant \frac{M_p l^2}{3EI}$$

$$\psi_3 \leqslant 0 \qquad \delta \geqslant \frac{M_p l^2}{6EI}.$$

Each of these three conditions sets a lower limit on the value of δ, the greatest of which corresponds to the condition $\psi_1 \leqslant 0$. It follows that the last hinge forms at section 1, so that at the point of collapse $\psi_1 = 0$. The deflection δ_c at the point of collapse is therefore

$$\delta_c = \frac{2M_p l^2}{3EI}.$$

The values of ψ_2 and ψ_3 at the point of collapse are obtained from Equations (5.22) and (5.23), with $\delta = \delta_c$, and are

$$(\psi_2)_c = \frac{M_p l}{2EI}$$

$$(\psi_3)_c = -\frac{M_p l}{2EI}.$$

The residual bending moment distributions (i) and (ii) cannot be uniquely chosen. Any two independent distributions which satisfy Equation (5.17) will suffice to establish two compatibility equations. The working will, of course, tend to be arithmetically easier if the distributions involve zeros. Similar remarks can be made about the distribution M^*.

The method just described could be used for more complex problems. However, it is possible to shorten the working slightly by assuming at the outset that a particular hinge is the last to form. Thus, in the above example, it might have been assumed (incorrectly) that the last hinge formed at section 3. It would then follow from Equations (5.18)–(5.20) that

$$\psi_1 = \frac{M_p l}{4EI}$$

$$\psi_2 = -\frac{M_p l}{4EI}$$

$$\psi_3 = 0$$

$$\delta = \frac{M_p l^2}{6EI}.$$

Both ψ_1 and ψ_2 are seen to have the wrong sign, so that the last hinge cannot be formed at section 3. The solution may be modified by adding the deflections and hinge rotations of the mechanism of Fig. 5.6(b), for compatibility is not thereby affected. This gives

$$\psi_1 = \frac{M_p l}{4EI} - \theta$$

$$\psi_2 = -\frac{M_p l}{4EI} + 3\theta$$

$$\psi_3 = -2\theta$$

$$\delta = \frac{M_p l^2}{6EI} + 2l\theta.$$

By inspection, $\theta = M_p l/4EI$ brings ψ_1 to zero and makes ψ_2 positive and ψ_3 negative, and with this value of θ the previous solution is obtained.

5.4.4 Rectangular frame

Consider the rectangular frame whose dimensions and loading are shown in Fig. 5.7(a). All the members of this frame are of the same uniform section, with flexural rigidity EI and plastic moment M_p. It is required to find the horizontal and vertical deflections h and v, which correspond to the loads H and V, at the point of collapse. The first case to be examined will be $H = V = W$. For this frame

$$n = 5$$

$$r = 3 = \text{number of independent equations of compatibility}$$

$$n - r = 2 = \text{number of independent equations of equilibrium.}$$

The two independent equations of equilibrium may be obtained by applying the virtual displacement method to the two independent mechanisms shown in Fig. 5.7(b) and (c), and are

$$-M_1 + M_2 - M_4 + M_5 = Hl. \tag{5.24}$$

$$-M_2 + 2M_3 - M_4 = Vl. \tag{5.25}$$

When $H = V = W$, collapse occurs by the combined mechanism shown in Fig. 5.7(d) when $W = W_c = 3M_p/l$. The coresponding bending moment distribution is given in Table 5.2.

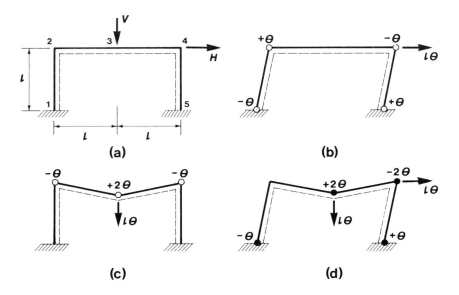

Fig. 5.7 *Fixed-base rectangular frame*
    (a) Dimensions and loading
    (b) and (c) Independent virtual mechanisms
    (d) Collapse mechanism

Three independent residual moment distributions m^* are needed to establish the three compatibility equations. These must satisfy the two equations of equilibrium for residual moments which follow at once from Equations (5.24) and (5.25), and are

$$-m_1 + m_2 - m_4 + m_5 = 0. \tag{5.26}$$

$$-m_2 + 2m_3 - m_4 = 0. \tag{5.27}$$

The particular distributions selected are shown in Table 5.2; they follow a pattern suggested by Heyman (1961) which is readily extended to multistorey, multibay frames.

Two distributions of bending moment M^* are required for the determination of the deflections h and v. These must satisfy respectively the requirements of equilibrium with the unit loads $H = 1$ (with $V = 0$) and $V = 1$ (with $H = 0$). These are derived from Equations (5.24) and (5.25), and are also shown in Table 5.2.

Using in turn the distributions (i)–(v) in Equations (5.13) and (5.15), as appropriate, the following equations result:

$$-\frac{M_p l}{12EI} + \psi_1 + 0.5\psi_3 = 0$$

$$\frac{M_p l}{12EI} + 0.5\psi_3 + \psi_4 + \psi_5 = 0$$

$$\frac{M_p l}{6EI} + \psi_3 + \psi_4 = 0$$

$$\frac{M_p l^2}{3EI} - l\psi_1 = h$$

$$\frac{M_p l^2}{4EI} + 0.5l\psi_3 = v.$$

ψ_2 is taken as zero, as there is no hinge at section 2 in the collapse mechanism.

Table 5.2 *Virtual force and actual displacement systems for frame of Fig. 5.7*

Section		1	2	3	4	5
Virtual force systems						
m^*	(i)	1	1	0.5	0	0
	(ii)	0	0	0.5	1	1
	(iii)	0	1	1	1	0
M^*	(iv)	$-l$	0	0	0	0
	(v)	0	0	0.5l	0	0
Actual displacement system						
$EI\kappa = M$		$-M_p$	0	M_p	$-M_p$	M_p
ϕ		ψ_1	ψ_2	ψ_3	ψ_4	ψ_5

Making the arbitrary assumption that $\psi_3 = 0$, these five equations are easily solved. The solution is given below; to each resulting hinge rotation and deflection there has been added a term for the motion of the collapse mechanism of Fig. 5.7(d).

$$\psi_1 = \frac{M_p l}{12EI} - \theta$$

$$\psi_3 = 2\theta$$

$$\psi_4 = -\frac{M_p l}{6EI} - 2\theta$$

$$\psi_5 = \frac{M_p l}{12EI} + \theta$$

$$h = v = \frac{M_p l^2}{4EI} + l\theta.$$

To conform with the signs of the plastic moments in the collapse mechanism, ψ_1 and ψ_4 must be negative and ψ_3 and ψ_5 must be positive. By inspection, this is first achieved when

$$\theta = \frac{M_p l}{12EI}.$$

With this value of θ, ψ_1 is zero and there is conformity of signs at the other three hinges. It is concluded that the last hinge forms at section 1, and it follows that at the point of collapse

$$(\psi_1)_c = 0$$

$$(\psi_3)_c = \frac{M_p l}{6EI}$$

$$(\psi_4)_c = -\frac{M_p l}{3EI}$$

$$(\psi_5)_c = \frac{M_p l}{6EI}$$

$$h_c = v_c = \frac{M_p l^2}{3EI}.$$

5.4.5 Partial collapse

In the example just considered the collapse mechanism was of the complete type, with $(r + 1)$ plastic hinges in a collapse mechanism which had only one degree of freedom. The entire frame was therefore statically determinate at collapse. When the collapse is partial, with fewer than $(r + 1)$ plastic hinges, the bending moment distribution at collapse cannot be fully determined by statics alone, but this does not give rise to any additional difficulty in determining deflections at the point of collapse.

Take as an example the frame of Fig. 5.7(a), with $V = W$ and $H = W/6$. Collapse now occurs by the beam mechanism of Fig. 5.7(c), which involves only three plastic hinges and is therefore partial. The bending moments at these hinges are

$$M_2 = -M_p, \quad M_3 = M_p, \quad M_4 = -M_p.$$

Substituting these values in the two equilibrium equations (5.24) and (5.25), it is found that

$$W = W_c = 4M_p/l$$

$$-M_1 + M_5 = 2M_p/3. \tag{5.28}$$

The separate values of M_1 and M_5 can only be found by invoking considerations other than those of equilibrium.

Three compatibility equations may now be established as in the previous example, using in turn the same distributions (i)–(iii) which were set out in Table 5.2. The penultimate line of this table, representing the actual bending moment distribution, is replaced by

Section	1	2	3	4	5
$EI\kappa = M$	M_1	$-M_p$	M_p	$-M_p$	M_5

Two equations for the determination of the deflections h and v can also be found by using the distributions (iv) and (v) in Table 5.2. The resulting equations, assuming that $\psi_1 = \psi_5 = 0$, are

$$\frac{(M_1 - M_p)l}{2EI} + \psi_2 + 0.5\psi_3 = 0. \tag{5.29}$$

$$\frac{(M_5 - M_p)l}{2EI} + 0.5\psi_3 + \psi_4 = 0. \tag{5.30}$$

$$\frac{(M_1 + M_5 - 4M_p)l}{6EI} + \psi_2 + \psi_3 + \psi_4 = 0. \tag{5.31}$$

$$\frac{(M_p - 2M_1)l^2}{6EI} = h. \tag{5.32}$$

$$\frac{M_p l^2}{6EI} + 0.5l\psi_3 = v. \tag{5.33}$$

Assuming arbitrarily that the last hinge forms at section 3, so that $\psi_3 = 0$, Equations (5.29) and (5.30) give

$$\psi_2 = \frac{(M_p - M_1)l}{2EI}, \tag{5.34}$$

$$\psi_4 = \frac{(M_p - M_5)l}{2EI}, \tag{5.35}$$

and substituting in Equation (5.31), it is found that

$$M_1 + M_5 = M_p. \tag{5.36}$$

Combining this result with the equilibrium equation (5.28),

$$M_1 = M_p/6, \quad M_5 = 5M_p/6.$$

Substituting these values in Equations (5.32)–(5.35), and adding terms for the motion of the mechanism of Fig. 5.7(c) gives

$$\psi_2 = \frac{5M_p l}{12EI} - \theta$$

$$\psi_3 = 2\theta$$

$$\psi_4 = \frac{M_p l}{12EI} - \theta$$

$$h = \frac{M_p l^2}{9EI}$$

$$v = \frac{M_p l^2}{6EI} + l\theta.$$

ψ_2 and ψ_4 must both be negative to conform with the signs of the plastic hinges at these sections, while ψ_3 must be positive. Inspection shows that the last hinge to form must be at section 2, with $\theta = 5M_p l/12EI$, so that at the point of collapse

$$(\psi_2)_c = 0$$

$$(\psi_3)_c = \frac{5M_p l}{6EI}$$

$$(\psi_4)_c = -\frac{M_p l}{3EI}$$

$$h = \frac{M_p l^2}{9EI}$$

$$v = \frac{7M_p l^2}{12EI}.$$

The feature of this analysis is that the three compatibility equations (5.29)–(5.31) furnish a relationship between M_1 and M_5, Equation (5.36), which when combined with the equilibrium equation (5.28) enables their separate values to be found. In general, if the number of plastic hinges in the collapse mechanism is $(r + 1 - q)$, the r compatibility equations will provide q relationships between the unknown bending moments, thus enabling the problem to be solved. Heyman (1961) gives an example of a four-storey frame for which $q = 3$.

Symonds (1952) compared deflection estimates of this kind with the tests of Baker and Heyman (1950) on miniature rectangular frames, and found that the agreement was generally satisfactory. Vickery (1961) used deflection estimates obtained in this way to study the changes in plastic collapse loads due to deformations, and Onat (1955) has also considered the effect of deformations on the immediate post-collapse behaviour. The effects of flexibility of joints and supports on the deflections developed at the point of collapse were studied using this technique by Neal (1960), who showed that within certain limits these deflections are unaffected by such factors. A valuable study of the relationship between this method of analysis, the determination of upper and lower bounds on plastic collapse loads, and elastic analysis was made by Munro (1965).

References

Baker, J.F., and Heyman, J. (1950), 'Tests on miniature portal frames', *Struct. Engr*, 28, 139.

Barrett, A.J. (1953), 'Unsymmetrical bending and bending combined with axial loading of a beam of rectangular cross-section into the plastic range', *J. R. Aero. Soc.*, 57, 503.

Cook, G. (1937), 'Some factors affecting the yield point in mild steel', *Trans. Inst. Engrs Shipb., Scot.*, 81, 371.

Dwight, J.B. (1953), 'An investigation into the plastic bending of aluminium alloy beams', Research Report no. 16., Aluminium Development Association.

Finzi, L. (1957), 'Unloading processes in elastic-plastic structures', 9th Int. Congr. Appl. Mech., Brussels, 1957.

Fritsche, J. (1930), 'Die Tragfähigkeit von Balken aus Stahl mit Berücksichtigung des plastischen Verformungsvermogens', *Bauingenieur*, 11, 851.

Heyman, J. (1961), 'On the estimation of deflexions in elastic-plastic framed structures', *Proc. Inst. Civil Engrs*, 19, 39.

Horne, M.R. (1948), Discussion of: 'Theory of inelastic bending with reference to limit design', *Trans. Am. Soc. Civil Engrs*, 113, 250.

Horne, M.R. (1951), 'Effect of strain-hardening on the equalisation of moments in the simple plastic theory', *Weld. Res.*, 5, 147.

Hrennikoff, A. (1948), 'Theory of inelastic bending with reference to limit design', *Trans. Am. Soc. Civil Engrs*, 113, 213.

Munro, J. (1965), 'The elastic and limit analysis of planar skeletal structures', *Civ. Eng. Publ. Wks Rev.*, 60, May.

Neal, B.G. (1960), 'Deflections of plane frames at the point of collapse', *Struct. Engr.*, 38, 224.

Onat, E.T. (1955), 'On certain second order effects in the limit design of frames', *J. Aero. Sci.*, 22, 681.

Rawlings, B. (1956), 'The analysis of partially plastic redundant steel frames', *Aust. J. Appl. Sci.*, 7, 10.

Roderick, J.W. (1954), 'The load-deflection relationship for a partially plastic rolled steel joist', *B. Weld. J.*, 1, 78.

Roderick, J.W. (1960), 'The elasto-plastic analysis of two experimental portal frames', *Struct. Engr*, 38, 245.

Roderick, J.W. and Heyman, J. (1951), 'Extension of the simple plastic theory to take account of the strain-hardening range', *Proc. Inst. Mech. Engrs*, 165, 189.

Roderick, J.W. and Phillipps, I.H. (1949), 'The carrying capacity of simply supported mild steel beams', *Research (Eng. Struct. Suppl.), Colston Papers*, **2**,
9.

Roderick, J.W. and Pratley, H.H.L. (1954), 'The behaviour of rolled steel joists
in the plastic range', *B. Weld. J.*, **1**, 261.

Sawyer, H.A. (1961), 'Post-elastic behaviour of wide flange steel beams', *J.
Struct. Div., Proc. Am. Soc. Civil Engrs*, **87** (ST 8), 43.

Symonds, P.S. (1952), Discussion of 'Plastic design and the deformation of
structures', *Weld. J., Easton, Pa.*, **31**, 33-s.

Symonds, P.S. and Neal, B.G. (1951), 'Recent progress in the plastic methods of
structural analysis', *J. Franklin. Inst.*, **252**, 383.

Symonds, P.S. and Neal, B.G. (1952), 'The interpretation of failure loads in the
plastic theory of continuous beams and frames', *J. Aero. Sci.*, **19**, 15.

Tanaka, H. (1961), 'A systematic calculation of elastic-plastic deformation of
frames at imminent collapse', Report, Inst. Indust. Sci., Tokyo Univ.

Vickery, B.J. (1961), 'The influence of deformations and strain-hardening on the
collapse load of rigid frame steel structures', *Civ. Eng. Trans., Inst. Engrs
Austr.*, **103**, September.

Young, B.W. and Dwight, J.B. (1971), 'Residual stresses and their effect on the
moment-curvature properties of structural steel sections', C.I.R.I.A. Tech.
Note 32.

Examples

1. A uniform, simply supported rectangular beam of span $2l$ is of ideal plastic
material and has the bending moment-curvature relation given in Equations (5.1)
and (5.2). It carries a uniformly distributed load W which is increased steadily
from zero. Yield first occurs at mid-span when $W = W_0$, and the central deflection is then $\delta_0 = 5\kappa_y l^2/12$. Find the central deflection when $W = 9W_0/8$, and
sketch the form of the plastic zones.

2. A uniform, fixed-ended beam of length $3l$ has flexural rigidity EI and plastic
moment M_p. It carries a concentrated load $2W$ at a distance l from one end and
another concentrated load W at a distance l from the other end. Estimate the
deflections corresponding to each load at the point of collapse.

3. For the pinned-base rectangular frame whose dimensions and loading are
shown in Fig. 5.5, collapse occurs when $H = 1.16M_p/l$ and $V = 2.84M_p/l$.
Estimate the deflection h which corresponds to the load H at the point of
collapse.

4. In a fixed-base rectangular frame ABCD the columns AB and CD are of height
$2l$ and l respectively. The base A is lower than the base D by a height l, so that
the beam BC, of length $2l$, is horizontal. All members of the frame are uniform
and of the same cross section, with flexural rigidity EI and plastic moment M_p.
The beam BC carries a central concentrated vertical load W, and a concentrated

horizontal load W is applied at C in the direction BC. Find the horizontal deflection of C at the point of collapse.

5. A fixed-base rectangular frame is of height l and span kl. All the members are of uniform section with flexural rigidity EI and plastic moment M_p. It is subjected to a central concentrated vertical load $3W$ and a concentrated horizontal load W at the top of one of the columns. Estimate the deflections corresponding to the two loads at the point of collapse (a) if $k = 3$, and (b) if $k = 2$. (Hint: show that in case (b) a plastic hinge must form at one of the bases, although collapse occurs by the beam mechanism.)

6. In a two-storey, single-bay rectangular frame ABCDEF the continuous columns ABC and FED are each of total length $2l$, and AB = BC = DE = EF = l. The upper and lower beams CD and BE each have a span $2l$. The frame is rigidly jointed and rigidly built-in at the bases A and F. All the members are uniform and have the same flexural rigidity EI and plastic moment M_p. The frame carries concentrated vertical loads W at the centres of the beams CD and BE, and also concentrated horizontal loads $0.9W$ at both D and E, acting in the directions CD and BE. Estimate the horizontal deflection of E at the point of collapse.

6 Factors Affecting the Plastic Moment

6.1 Introduction

The plastic moment M_p has so far been assumed to have a constant value for a given member. This assumption will now be examined. The factors which affect the plastic moment fall into two categories. In the first place, it will be recalled that according to the simple theory set out in Chapter 1, $M_p = Z_p \sigma_0$, so that any factors which affect the yield stress σ_0 will equally affect the plastic moment M_p. These factors are discussed in Section 6.2. Second, members in a frame are generally called upon to resist not only bending moments, but also normal and shear forces. However, the fully plastic stress distribution which leads to the derivation of the value $Z_p \sigma_0$ for the plastic moment only produces a resultant bending moment, with zero resultant normal and shear forces. More complex distributions of both normal and shear stresses are therefore required, and the effect is to reduce the value of the plastic moment below $Z_p \sigma_0$. In most practical cases the reduction is not very large, and methods for making appropriate allowances are given in Sections 6.3 and 6.4.

Plastic hinges often occur beneath concentrated loads, and the plastic moment is then modified in addition by contact stresses. Again the effect is usually small; it can be allowed for on a semi-empirical basis, as discussed in Section 6.5.

6.2 Variations in yield stress

As pointed out in Chapter 1, the yield stress of steel is strongly dependent on metallurgical factors, namely the chemical composition and heat treatment. These factors will not be discussed here; rather, the concern is with the ways in which the yield stress of a given steel is affected by those environmental influences to which it may be exposed during normal structural use.

The first factor to be considered is the rate of loading. Many investigators have shown that in tensile tests the yield stress of mild steel is affected by the strain rate, and a comprehensive literature review has been given by Mainstone (1975). To give one example, a comprehensive series of tests on three steels was carried out by Rao, Lohrmann and Tall (1966). In these tests a static yield stress was determined at effectively zero strain rate, and dynamic yield stresses were also measured at rates of strain up to about 1.5×10^{-3}/s. This strain rate is such that the yield point is reached in about 1 s, and thus represents a fairly rapid rate of loading. It was found that for A.S.T.M. structural steel (A36-63T) the yield stress was increased by about 13 per cent at the highest strain rate used in the tests.

Rates of loading normally encountered would not produce such high strain rates, and so variations in the yield stress due to this effect are likely to be of smaller order, say 1–2 per cent. However, it should be noted that impact loading may well result in even higher strain rates and so cause a substantial enhancement of yield stress.

The lower yield stress of mild steel may be affected by strain-ageing. If a tensile specimen is subjected to a load which causes some degree of plastic strain, it is found that after a fairly considerable length of time at ambient temperature the yield point for a further loading in the same sense is increased (Baird, 1963). This phenomenon of strain-ageing is accelerated considerably if the temperature is raised, and this is exploited in laboratory tests. It should not be confused with the increase in the elastic limit which occurs after the material has entered the strain-hardening range.

Strain-ageing effects will occur in a framed structure after it has been subjected to loads which cause the formation of one or more plastic hinges. This might well happen in practice; there is no reason why some plastic hinges should not form when, for example, the structure experiences the characteristic loads (see Section 2.7), even though the structure is not then dangerously near a plastic collapse state. After a lapse of several months, strain-ageing at these hinge positions would cause an increase of yield stress, and thus of the plastic moment, of the order of 10 per cent. However, unless the loading in question was close to the plastic collapse load, the number of hinge positions concerned would only constitute a small fraction of the total involved in the collapse mechanism. The collapse load factor λ_c would therefore not be greatly affected. Moreover, the effect would be to increase λ_c, and is therefore on the safe side.

It is difficult to determine the plastic moment accurately by finding the yield stress σ_0 from a tensile test and then computing the plastic moment as $Z_p \sigma_0$. In flexure the strain, and therefore the rate of strain, varies across the section in a roughly linear manner. Since σ_0 depends on the strain rate, there will be some variation in its value across the section. Moreover, constant rates of strain rarely occur in either full-scale or model structures. Increments of load cause initially rapid rates of deflection followed by a progressively slower approach towards final equilibrium. The specification of the strain rate at which a tensile test should be conducted to allow the prediction of the plastic moment is therefore almost purely empirical.

A further problem arises in the case of rolled steel sections, in which the yield stress varies quite widely between specimens cut from different locations. These variations, due to differences in amounts of plastic working and in rates of cooling during the rolling process, may be as large as $60 \, \text{N/mm}^2$, as pointed out by Baker (1972). It may be concluded that the most reliable way of determining the plastic moment of a beam is by means of a bending test.

6.3 Effect of normal force

The effect on the plastic moment of a normal force will be considered for a beam whose cross section has two axes of symmetry, with the plane of bending coinciding with one of these axes, as in Fig. 6.1(a). It will be assumed that the neutral surface is plane and perpendicular to the plane of bending.

Suppose that there is a tensile normal force N, and that at full plasticity the moment, which would have been M_p in the absence of normal force, is reduced to M_N. The stress distribution will be as shown in Fig. 6.1(b), the neutral axis being displaced downwards by a distance e from the relevant axis of symmetry. Initially, e will be presumed to be less than $D/2$, so that the neutral axis remains within the cross section.

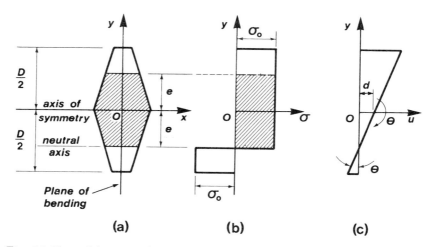

Fig. 6.1 *Normal force combined with bending*
(a) Cross section
(b) Fully plastic stress distribution; $e < D/2$
(c) Normal displacements

The shaded area of the cross section between $y = \pm e$ is subjected to a uniform tensile stress σ_0. This part of the stress distribution therefore has a resultant which is a normal force through the centroid O of the cross section. The two unshaded areas are subjected to uniform stresses of the same magnitude σ_0 but of opposite sign, whose resultant must be a pure couple about the axis of symmetry Ox. The shaded area of the cross section, denoted by A_e, can therefore be regarded as carrying the normal force N, so that

$$N = A_e \sigma_0. \tag{6.1}$$

The value of the normal force which would cause full plasticity in the absence of any bending moment is defined as N_p, the plastic thrust or squash load. If the total area of the cross section is A,

$$N_p = A\sigma_0 \qquad (6.2)$$

so that

$$n = \frac{N}{N_p} = \frac{A_e}{A}. \qquad (6.3)$$

The effect of N is to reduce the plastic moment by the amount $(M_p)_e$ which would be contributed by the shaded area in the absence of any normal force,

$$M_N = M_p - (M_p)_e. \qquad (6.4)$$

Defining $(M_p)_e = (Z_p)_e \sigma_0$, and recalling that $M_p = Z_p \sigma_0$,

$$M_N = M_p \left[1 - \frac{(Z_p)_e}{Z_p} \right]. \qquad (6.5)$$

This result is also valid for a compressive normal force because of the double symmetry.

Fig. 6.1(c) shows the normal displacements u at the plastic hinge. A rotation θ takes place about the neutral axis, which is a distance e below the axis of symmetry Ox. Thus, at the axis of symmetry there is an axial displacement d corresponding to N in addition to the rotation θ corresponding to M_N, where

$$d = e\theta. \qquad (6.6)$$

The hinge may therefore be termed complex, and the work absorbed W is

$$W = M_N \theta + Nd. \qquad (6.7)$$

The above analysis assumed that e was less than $D/2$, so that the neutral axis remained within the cross section. Fig. 6.2(a) shows the distribution of normal displacements u when e exceeds $D/2$. These displacements are all tensile, so that the distribution of normal stress is as shown in Fig. 6.2(b). The normal force is now the squash load N_p, and the moment is zero, so that the work absorbed is simply

$$W = N_p d. \qquad (6.8)$$

However, at the axis of symmetry there is a rotation θ in addition to the axial displacement d corresponding to N_p, and Equation (6.6) still holds. Since $e \geqslant D/2$, it follows that

$$\theta \leqslant \frac{2d}{D}, \qquad (6.9)$$

and so despite the fact that the moment is zero the hinge is still complex, with a rotation θ of any magnitude consistent with Equation (6.9).

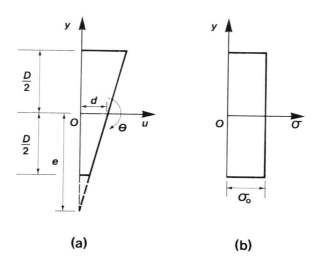

(a)   **(b)**

Fig. 6.2 *Normal force combined with bending*
(a) Normal displacements; $e > D/2$
(b) Fully plastic stress distribution

The relations (6.6)–(6.9) should strictly be used in establishing the work equation corresponding to a possible collapse mechanism, as pointed out by Heyman (1975). However, as will be seen in Section 6.3.3, the effect of normal force is often negligibly small in practical cases.

Cases in which the cross section has only one axis of symmetry, which coincides with the plane of bending, were treated by Eickhoff (1954). It is important to specify the axis about which the resultant moment is measured, and Eickhoff chose for this purpose the equal area axis.

6.3.1 *Rectangular cross section*

For the rectangular cross section shown in Fig. 6.3(a),

$$A = BD, \qquad A_e = 2Be$$

$$Z_p = \tfrac{1}{4}BD^2, \quad (Z_p)_e = Be^2,$$

so that in Equations (6.3) and (6.5),

$$n = \frac{2e}{D}$$

$$M_N = M_p \left[1 - \frac{4e^2}{D^2}\right] = M_p(1 - n^2). \tag{6.10}$$

This interaction relation between M_N/M_p and n was first obtained by Girkmann (1931).

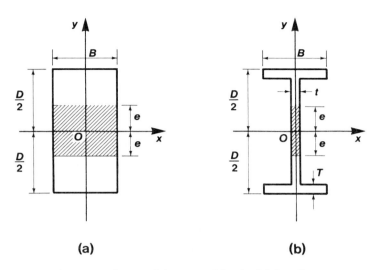

(a)   **(b)**

Fig. 6.3 *Particular cases of normal force combined with bending*
(a) Rectangular cross section
(b) I-section bent about major axis

Shakir-Khalil and Tadros (1973) considered rectangular sections subjected to bending about both axes of symmetry in the presence of axial force. Both solid and hollow sections were covered, and the results were verified experimentally.

6.3.2 *I-section bent about major axis*

An I-section may be regarded as consisting of three rectangles, as in Fig. 6.3(b), without much loss of accuracy. If the neutral axis lies in the web, as shown, the value of A_e is $2et$, so that from Equation (6.3)

$$n = \frac{2et}{A}, \tag{6.11}$$

where A is the total area of the cross section. This relation holds true so long as A_e does not exceed the web area $A_w = t(D - 2T)$, so that

$$n \leqslant \frac{A_w}{A}.$$

The value of $(Z_p)_e$ is that appropriate for a rectangular section of breadth t and depth $2e$, so that

$$(Z_p)_e = te^2$$

$$= \frac{n^2 A^2}{4t}, \tag{6.12}$$

making use of Equation (6.11). Substituting in Equation (6.5),

$$M_N = M_p(1 - kn^2); \qquad k = \frac{A^2}{4tZ_p}$$

for

$$n \leqslant \frac{A_w}{A}. \tag{6.13}$$

The plastic section modulus Z_p was shown in Section 1.4, Equation (1.14), to be

$$Z_p = BT(D - T) + \tfrac{1}{4}t(D - 2T)^2.$$

Using this result, it can be shown that

$$\frac{1}{k} = 1 - \left(\frac{A_f}{A}\right)^2 \left(1 - \frac{t}{B}\right), \tag{6.14}$$

where A_f is the total flange area $2BT$.

If the neutral axis lies in a flange, a similar analysis gives the result

$$M_N = M_p k \frac{t}{B}(1 - n)\left[\frac{2BD}{A} - 1 + n\right]$$

for

$$\frac{A_w}{A} \leqslant n \leqslant 1. \tag{6.15}$$

Results equivalent to Equations (6.13)–(6.15) were first obtained by Girkmann (1931). These equations form the basis of the expressions for the reduction of plastic section modulus which are published in section tables. Tests on columns under combined thrust and moment were carried out by Beedle, Ready and Johnston (1950), and good agreement with this theory was obtained.

The interaction relation between M_N/M_p and n for a typical I-section is shown in Fig. 6.4, which also gives the parabolic interaction relation of Equation (6.10) for a rectangular section.

6.3.3 Effect of normal force in practical cases

For a beam of I-section bent about the major axis, the reduction of plastic moment due to normal force is less than 2 per cent when $n = 0.1$. In single-storey frames it is found that the value of n is usually less than 0.1, so that the effect of normal force will be small, unless the frame has some special feature such as crane-bearing columns. However, in multistorey frames an allowance for the effect of normal force will often have to be made for columns in the lower storeys, and the effect of normal force on the plastic moment is likely to be of importance in shallow arches.

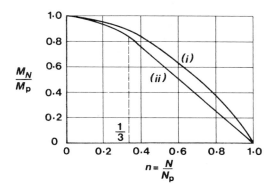

Fig. 6.4 *Effect of normal force on plastic moment*

(i) Rectangular cross section
(ii) I-section bent about major axis: $A_f = \frac{2}{3}A$, $t = 0.063B$. Neutral axis in flange when n exceeds $1/3$

6.4 Effect of shear force

As pointed out by Drucker (1956), there is no reason to expect that a unique interaction relation exists between plastic moment and shear force for a given cross section. The studies which have been made of this problem have usually been based on a cantilever of length L carrying a concentrated transverse load F at its free end. A cantilever of length $2L$ carrying a uniformly distributed load F would have the same *local* situation at the clamped end of a shear force F and a bending moment FL. However, the collapse value of F must in each case be derived by considering conditions throughout the full length of the cantilever, and there is no *a priori* reason for supposing that the same collapse load would be found in both cases.

Despite this, it is common practice to present the results of shear force analyses as an interaction relation between shear force and plastic moment. This cannot be strictly justified, but fortunately the effects of shear force are generally small in practical cases. It is therefore not wholly unreasonable to use an interaction relation to estimate the reduction of plastic moment due to shear, even when the loading differs from that which forms the basis of the interaction relation.

Several analyses of the effect of shear force have been concerned with stress distributions or plastic deformations at the critical section only, no regard being paid to conditions in the remainder of the cantilever. Such local analyses are unreliable, since as pointed out by Drucker (1956) neither a lower nor an upper bound is obtained.

6.4.1 *Rectangular cross section*

Consider a cantilever of rectangular cross section, breadth B and depth D, which is subjected to an end shear load F, as shown in Fig. 6.5. The length of the cantilever is L, so that the bending moment at the clamped end is FL. It is required to determine the value M_F of this moment at which full plasticity occurs, and in particular to compare its value with the plastic moment $M_p = \frac{1}{4}BD^2\sigma_0$ in the absence of shear.

A simple lower bound approach will now be outlined. Fig. 6.5 depicts a situation in which the cantilever is wholly elastic to the left of section AA. At this section the longitudinal normal stress σ_x just attains the yield value σ_0 in the outermost fibres. To the right of section AA yield zones are assumed to form in accordance with the simple theory of flexure given in Section 1.3.1, in which the effects of shear were not considered. In these yield zones the state of stress is

$$\sigma_x = \pm\sigma_0, \quad \sigma_y = 0, \quad \tau = 0,$$

the notation for stresses being as shown in the figure. Justification for this assumption was provided by Prager and Hodge (1951) and also by Horne (1951).

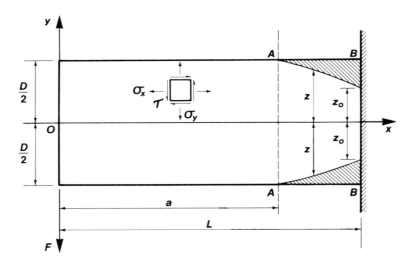

Fig. 6.5 *Plastic zones in cantilever of rectangular cross section*

Since τ is assumed to be zero in the yield zones, the shear force F must be carried entirely by shear stresses in the elastic core of total depth $2z$. As z

decreases the magnitudes of these shear stresses must increase, so that at some section BB yield must occur within the core. The analysis which follows is aimed at determining statically admissible stress distributions throughout the beam which are also safe, so that the yield criterion is not violated, but is just met at the critical section BB.

The von Mises yield criterion will be used. For the condition of plane stress which is assumed, this takes the form

$$\phi = [\sigma_x^2 - \sigma_x\sigma_y + \sigma_y^2 + 3\tau^2]^{1/2} \leqslant \sigma_0. \tag{6.16}$$

The yield stress in pure shear is $\sigma_0/\sqrt{3}$ according to this criterion.

The two equations of equilibrium are

$$\frac{\partial \sigma_x}{\partial x} + \frac{\partial \tau}{\partial y} = 0. \tag{6.17}$$

$$\frac{\partial \sigma_y}{\partial y} + \frac{\partial \tau}{\partial x} = 0. \tag{6.18}$$

In the elastic portion of the beam, $x \leqslant a$, σ_y is assumed to be zero, while σ_x varies linearly across the section in accordance with the simple theory of elastic flexure, so that

$$\sigma_x = \frac{12\,Fxy}{BD^3} \tag{6.19}$$

$$\sigma_y = 0. \tag{6.20}$$

The corresponding solution of Equations (6.17) and (6.18), bearing in mind that $\tau = 0$ when $y = \pm D/2$, is

$$\tau = \frac{3F}{2BD}\left[1 - \left(\frac{2y}{D}\right)^2\right]. \tag{6.21}$$

At section AA, where $x = a$, $\sigma_x = \sigma_0$ when $y = D/2$, so that from Equation (6.19)

$$\sigma_0 = \frac{6Fa}{BD^2},$$

$$Fa = M_y = \tfrac{1}{6}BD^2\sigma_0. \tag{6.22}$$

The distributions of the stresses σ_x and τ across section AA are shown in Fig. 6.6. It will be assumed in what follows that when $x \leqslant a$, τ nowhere exceeds the yield stress in pure shear, $\sigma_0/\sqrt{3}$, so that from Equation (6.21)

$$\frac{3F}{2BD} \leqslant \frac{\sigma_0}{\sqrt{3}}$$

$$F \leqslant \frac{2}{3\sqrt{3}} BD\sigma_0. \tag{6.23}$$

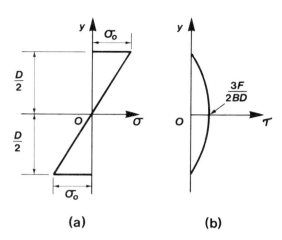

(a)   **(b)**

Fig. 6.6 *Stress distributions in cantilever of rectangular cross section*
(a) Distribution of σ_x at section AA
(b) Distribution of τ at section AA

If the plastic shear force F_p is defined as the shear force corresponding to the development of the yield stress in pure shear over the entire cross section,

$$F_p = \frac{1}{\sqrt{3}} BD\sigma_0, \tag{6.24}$$

so that condition (6.23) becomes

$$f = \frac{F}{F_p} \leqslant \frac{2}{3}. \tag{6.25}$$

In the limiting case when condition (6.23) becomes an equality, the situation at section AA is

$$\sigma_x = \sigma_0\left(\frac{2y}{D}\right)$$

$$\tau = \frac{1}{\sqrt{3}}\sigma_0\left[1 - \left(\frac{2y}{D}\right)^2\right],$$

and substituting in the yield criterion (6.16)

$$\phi = \sigma_0\left[1 - \left(\frac{2y}{D}\right)^2 + \left(\frac{2y}{D}\right)^4\right]^{1/2}.$$

It is readily shown that $\phi \leqslant \sigma_0$ for $-D/2 \leqslant y \leqslant D/2$, so that this distribution of stress is safe.

Turning now to the elastic-plastic portion of the beam, $x \geqslant a$, it is assumed that in the elastic core σ_x still varies linearly with y, so that

$$\sigma_x = \sigma_0 \left(\frac{y}{z} \right). \tag{6.26}$$

In the plastic zones $\sigma_x = \pm\sigma_0$, while τ and σ_y are both zero. It follows from the simple theory of elastic-plastic flexure presented in Section 1.3.1, Equation (1.2), that the bending moment M is given by

$$M = Fx = B \left[\frac{D^2}{4} - \frac{1}{3} z^2 \right] \sigma_0, \tag{6.27}$$

so that

$$\frac{dz}{dx} = -\frac{3F}{2Bz\sigma_0}.$$

Using this result and Equation (6.26), the solution of the equilibrium Equations (6.17) and (6.18) is

$$\tau = \frac{3F}{4Bz} \left[1 - \left(\frac{y}{z} \right)^2 \right]. \tag{6.28}$$

$$\sigma_y = -\frac{9F^2}{8B^2 z^2 \sigma_0} \left(\frac{y}{z} \right) \left[1 - \left(\frac{y}{z} \right)^2 \right]. \tag{6.29}$$

The distributions of stress defined by Equations (6.26), (6.28) and (6.29) are statically admissible. It is now assumed that a critical condition is reached at section BB, where $z = z_0$, due to τ attaining the value $\sigma_0/\sqrt{3}$ at $y = 0$, so that

$$\frac{3F}{4Bz_0} = \frac{\sigma_0}{\sqrt{3}}. \tag{6.30}$$

The distributions of stress at this section then become

$$\sigma_x = \sigma_0 \left(\frac{y}{z_0} \right)$$

$$\tau = \frac{\sigma_0}{\sqrt{3}} \left[1 - \left(\frac{y}{z_0} \right)^2 \right]$$

$$\sigma_y = -\frac{2}{3} \sigma_0 \left(\frac{y}{z_0} \right) \left[1 - \left(\frac{y}{z_0} \right)^2 \right]. \tag{6.31}$$

These distributions are shown in Fig. 6.7. Substituting in the yield criterion, Equation (6.16),

$$\phi = \left[1 + \frac{1}{9}\left(\frac{y}{z_0}\right)^2 \left[1 - \left(\frac{y}{z_0}\right)^2 \right]\left[1 - 4\left(\frac{y}{z_0}\right)^2 \right]\right]^{1/2} \sigma_0.$$

Within the range $-z_0 \leqslant y \leqslant z_0$, ϕ is found to have a maximum value $1.003\sigma_0$ when $y = \pm 0.34 z_0$, so that the yield criterion is just not met. However, since the extent of the violation is only a maximum of 0.3 per cent, the stress distributions will be accepted as safe as well as being statically admissible. It is readily verified that BB is indeed the most critical section, as is to be expected.

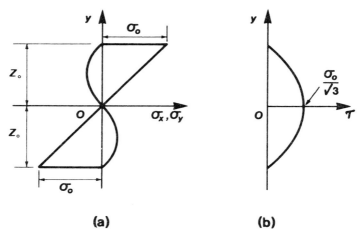

(a)   **(b)**

Fig. 6.7 *Stress distributions in cantilever of rectangular cross section*
(a) Distributions of σ_x and σ_y at section BB
(b) Distribution of τ at section BB

The value of M_F is found from Equation (6.27), with z having the value z_0 defined by Equation (6.30). Using Equation (6.24),

$$M_F = M_p \left[1 - 0.75 f^2 \right]. \tag{6.32}$$

This is an interaction relation between M_F/M_p and f. It is subject to the limitation set by condition (6.25), namely that f cannot exceed 2/3. When f has this value, Equation (6.32) shows that $M_F = 2M_p/3 = M_y$, and from Equation (6.30) it is found that $z_0 = D/2$. This means that the sections AA and BB shown in Fig. 6.5 coincide, and the length of the beam is then a, which is found from Equation (6.22) to be $D\sqrt{3}/4 = 0.433D$. The condition $f \leqslant 2/3$ is therefore equivalent to $L/D \geqslant 0.433$. Since the plastic theory is scarcely applicable to such short beams, this condition is not restrictive.

A more detailed study was made by Horne (1951), in which the growth of central yield zones in the neighbourhood of section BB was considered. This gave results similar in form to those just obtained, differing only in the values of the numerical constants, as follows:

$$M_F = M_p [1 - 0.44 f^2]; \quad f \leqslant 0.79. \tag{6.33}$$

This can be compared with the recommendation of Drucker (1956), who developed a lower bound solution based on more hypothetical stress distributions, and also discussed various upper bound approaches. The interaction relation proposed by Drucker is

$$M_F = M_p [1 - f^4]; \quad f \leqslant 1. \tag{6.34}$$

Green (1954) gave upper bound solutions, based on a consideration of deformations, for the cases of plane stress and plane strain. In this paper the influence of the nature of the support conditions was also discussed. The most notable feature of Green's solutions is that they show an increase in the plastic moment with shear force for small values of f. Onat and Shield (1954) have also given an upper bound solution for the plane strain problem.

6.4.2 I-section bent about major axis

For an I-section bent about its major axis it will be assumed that the shear force is carried solely by the web. Thus if the area of the web is A_w, the plastic shear force F_p is given by

$$F_p = \frac{1}{\sqrt{3}} A_w \sigma_0 \tag{6.35}$$

if the von Mises yield criterion is used. (If the Tresca yield criterion is used, the factor $1/\sqrt{3}$ in Equation (6.35) is replaced by $1/2$.) The upper bound solution of Leth (1954) showed that F can only exceed F_p for beams whose length/depth ratio is so small that the concept of beam action is meaningless, which verifies this assumption.

The plastic moment of an I-section bent about its major axis in the absence of shear was evaluated in Section 1.4. Assuming the section to consist of a pair of flanges, each of breadth B and thickness T, and a web of depth $(D - 2T)$ and thickness t, it was shown, Equation (1.14), that

$$M_p = [BT(D - T) + \tfrac{1}{4} t (D - 2T)^2] \sigma_0.$$

Neglecting T in comparison with D, this simplifies to

$$M_p = [BDT + \tfrac{1}{4} t D^2] \sigma_0$$
$$= \tfrac{1}{4} D [2A_f + A_w] \sigma_0. \tag{6.36}$$

In this equation the first term represents the contribution of the flanges, of total area A_f, and the second term the contribution of the web, of area A_w.

The simplest approach to the problem is to regard the flange moment as unaffected by shear and to modify the web moment in the same way as for a rectangular section. This method was adopted by Horne (1951); using his result as expressed in Equation (6.33) gives

$$M_F = \tfrac{1}{2}DA_f\sigma_0 + \tfrac{1}{4}DA_w\sigma_0(1-0.44f^2)$$

$$= M_p\left[1-0.44\left(\frac{A_w}{2A_f+A_w}\right)f^2\right]; \qquad f \leqslant 0.79. \qquad (6.37)$$

Leth (1954) showed that this procedure gives rise to stresses in the neighbourhood of the flange-web junction which violate the yield criterion, and gave suitable modifications to the theory. However, in both these analyses the conditions of equilibrium at the flange-web junction were not met, so that the solutions were not true lower bounds. A lower bound solution in which all the conditions of equilibrium are met and the yield criterion is not violated has been given by Neal (1961a). The results cannot be expressed in explicit form, but the interaction relation for a section with $A_f = 3A_w$ is shown in Fig. 6.8.

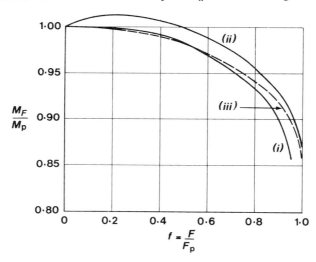

Fig. 6.8 *Effect of shear force on plastic moment of* I-*section bent about major axis:* $A_f = 3A_w$
  (i) Lower bound (Neal)
 (ii) Upper bound (Green)
(iii) Empirical relation (Heyman and Dutton)

Green (1954) applied his upper bound solution for a beam of rectangular cross section in plane stress to the case of a cantilever of I-section, assuming that the flange moment is unaffected by shear. The interaction relation for the case $A_f = 3A_w$ is also shown in Fig. 6.8.

It will be seen that the difference between the upper and lower bound solutions is never very large. An empirical relation close to these two bounds will therefore suffice for practical purposes. Such a relation was suggested by Heyman and Dutton (1954). It was derived from a purely local analysis of conditions at the critical section, so that its justification depends entirely upon its good agreement with the bounds illustrated in Fig. 6.8. The relation is

$$M_F = M_p \left[1 - \left(\frac{A_w}{2A_f + A_w} \right) \left(1 - (1 - f^2)^{1/2} \right) \right] ; \qquad f \leqslant 1. \quad (6.38)$$

It is suggested that this equation be used in practice. However, further work in this field is required. Progress is most likely to be made by the development of true upper and lower bound solutions. These are needed especially for the investigation of different loading arrangements to see whether or not it is really satisfactory to use, for a given cross section, a unique interaction relation such as Equation (6.38). Reference should be made to an upper bound analysis of a panel of a plate girder, in which a tension field develops in the web, described by Calladine (1973).

Comparison with experimental results is difficult, because until F approaches F_p, the effect of shear is quite small for I-sections, whereas when F is close to F_p, there is often no clearly defined collapse load. In their tests on model simply supported beams of I-section, Baker and Roderick (1940) and Hendry (1949, 1950) found this difficulty. However, in testing simply supported model plate girders subjected to central concentrated loads, Heyman and Dutton (1954) observed sharply defined collapse loads even when F was close to F_p, and their results agreed well with Equation (6.38). Further tests on plate girders by Longbottom and Heyman (1956), on model cantilevers of I-section by Green and Hundy (1957), and on a full-scale Vierendeel girder by Bull (1955), have provided additional confirmation of the usefulness of this empirical relation.

6.4.3 *Combined effect of shear and axial forces*

The combined effect of shear and axial forces acting simultaneously has not so far been discussed. However, only rarely will the shear and axial forces at a plastic hinge both be of sufficient magnitude to have an appreciable effect on the plastic moment. The problem was first discussed by Green (1954), who developed an upper bound solution for a beam of rectangular section in plane strain. Lower bound solutions were developed by Neal (1961b, 1961c) for cantilevers of rectangular and I-section, and Klöppel and Yamada (1958) gave a lower bound type of analysis for an I-section which is essentially a local criterion. Horne (1958) extended Heyman and Dutton's local analysis for an I-section to include the effect of axial force, and Kusuda and Thürlimann (1958) have also suggested similar approaches.

6.5 Contact stresses beneath loads

Model tests are often carried out on simply supported beams of rectangular cross section, subjected to either a central concentrated load or symmetrical two-point loading. For beams of similar section and material, the plastic moment is usually observed to be somewhat higher when central concentrated loading is imposed. The reason for this is that with symmetrical two-point loading the plastic moment is attained in the central length of the beam between the loads, which is in pure bending, so that the stresses in this region are only required to equilibrate a pure couple. This was the assumption made in the simple theory of Section 1.4, which led to the value $Z_p \sigma_0$ for the plastic moment. However, with central concentrated loading the plastic moment is only attained beneath the load, where the stresses must in addition balance shear forces and also the contact stresses due to the load. The effect is, paradoxically, to increase the collapse load and hence the value of the plastic moment derived from such a test.

An approximate elastic solution for the stresses in a simply supported rectangular beam subjected to a central concentrated load was developed by Stokes and published in a paper by Carus Wilson (1891). This solution indicates that the local effect due to the concentrated load consists principally of two stress components. The first of these is the compressive stress acting on planes parallel to the axis of the beam which is obviously required to support the load. The other is a longitudinal normal stress which modifies the linear distribution across the section obtained by the ordinary Bernoulli-Euler theory of elastic flexure. The effect of this latter stress is to multiply the ordinary bending stresses by a factor $(1 - kD/L)$, where D and L are the depth and length of the beam, respectively. k is a positive factor which varies across the section and is small except within a distance of the order of D on either side of the load, where it is of the order of unity.

This elastic solution supports the view of Roderick and Phillipps (1949) that the contact stresses tend to annul the ordinary bending stresses near the load, even when yield has occurred. Since the elastic solution suggests that the disturbance due to a central concentrated load is confined to a total length of about D, Roderick and Phillipps suggested that collapse will occur when the bending moment reaches the value $Z_p \sigma_0$ at a distance $D/2$ on either side of the load. The central bending moment at collapse is then $Z_p \sigma_0 L/(L - D)$, so that the apparent plastic moment M_p' is approximately

$$M_p' = Z_p \sigma_0 \left(1 + \frac{D}{L}\right). \tag{6.39}$$

Roderick and Phillipps carried out tests on a series of beams of rectangular cross section which confirmed this semi-empirical result. Further evidence in support was furnished by Heyman (1952), who used a relaxation method to determine the stress distribution in the neighbourhood of a central concentrated load on a beam of rectangular cross section at the point of collapse. The photo-elastic tests of Baes (1948) on beams of rectangular section and of Hendry (1949) on beams of I-section, also indicated that Equation (6.39) may be expected to be reasonably correct in many cases. For beams of I-section, the equation is clearly not applicable if bearing stiffeners are provided beneath the concentrated loads.

References

Baes, L. (1948), 'Les palplanches plates beval P pour constructions cellulaires', *Ossat. métall.*, **13**, 75.

Baird, J. D. (1963), 'Strain-ageing of steel — a critical review', *Iron and Steel*, **36**, 186.

Baker, J. F. and Roderick, J. W. (1940), 'Further tests on beams and portals', *Trans. Inst. Weld.*, **3**, 83.

Baker, M. J. (1972), 'Variability in the strength of structural steels — a study in structural safety: Part 1. Material variability', C.I.R.I.A. Tech. Note 44, September.

Beedle, L. S., Ready, J. A. and Johnston, B. G. (1950), 'Tests of columns under combined thrust and moment', *Proc. Soc. Exp. Stress Anal.*, **8**, 109.

Bull, F. B. (1955), 'Tests to destruction on a Vierendeel girder', prelim. vol., Conf. Correlation between Calculated and Observed Stresses and Displacements in Structures, Inst. Civil Engrs, 135.

Calladine, C. R. (1973), 'A plastic theory for collapse of plate girders under combined shearing force and bending moment', *Struct. Engr*, **51**, 147.

Drucker, D. C. (1956), 'The effect of shear on the plastic bending of beams', *J. Appl. Mech.*, **23**, 509.

Eickhoff, K. G. (1954), 'The plastic behaviour of sections having one axis of symmetry', B. Weld. Res. Assoc. Report FE. 1/37.

Girkmann, K. (1931), 'Bemessung von Rahmentragwerken unter Zugrundelegung eines ideal-plastischen Stahles', *S.B. Akad. Wiss. Wien* (Abt. IIa), **140**, 679.

Green, A. P. (1954), 'A theory of the plastic yielding due to bending of cantilevers and fixed-ended beams', *J. Mech. Phys. Solids*, **3**, 1, 143.

Green, A. P. and Hundy, B. B. (1957), 'Plastic yielding of I-beams: shear loading effects analysed', *Engineering*, **184**, 47.

Hendry, A. W. (1949), 'The stress distribution in a simply supported beam of I-section carrying a central concentrated load', *Proc. Soc. Exp. Stress Anal.*, **7**, 91.

Hendry, A. W. (1950), 'An investigation of the strength of certain welded portal frames in relation to the plastic method of design', *Struct. Engr*, **28**, 311.

Heyman, J. (1952), 'Elasto-plastic stresses in transversely loaded beams', *Engineering*, **173**, 359, 389.

Heyman, J. (1975), 'Overcomplete mechanisms of plastic collapse', *J. Optim. Theory Applic.*, **15**, 27.

Heyman, J. and Dutton, V. L. (1954), 'Plastic design of plate girders with unstiffened webs', *Welding and Metal Fabrication*, **22**, 265.

Horne, M. R. (1951), 'The plastic theory of bending of mild steel beams with particular reference to the effect of shear forces', *Proc. R. Soc., A.*, **207**, 216.

Horne, M. R. (1958), 'Full plastic moments of sections subjected to shear force and axial load', *B. Weld. J.*, **5**, 170.

Klöppel, K. and Yamada, M. (1958), 'Fliesspolyeden des Rechteck- und I-Quer-schnitte unter die Wirkung von Biegemoment Normalkraft und Querkraft', *Stahlbau*, **27**, 284.

Kusuda, T. and Thürlimann, B. (1958), 'Strength of wide flange beams under combined influence of moment, shear and axial force', Fritz. Eng. Lab. Report no. 248.1.

Leth, C-F. A. (1954), 'The effect of shear stresses on the carrying capacity of I-beams', Tech. Rep. A11-107, Brown Univ.

Longbottom, E. and Heyman, J. (1956), 'Tests on full-size and on model plate girders', *Proc. Inst. Civil Engrs*, **5**, (part III), 462.

Mainstone, R. J. (1975), 'Properties of materials at high rates of straining or loading', *Matériaux et Construction*, **8**, 102.

Neal, B. G. (1961a), 'Effect of shear force on the fully plastic moment of an I-beam', *J. Mech. Eng. Sci.*, **3**, 258.

Neal, B. G. (1961b), 'The effect of shear and normal forces on the fully plastic moment of a beam of rectangular cross section', *J. Appl. Mech.*, **28**, 269.

Neal, B. G. (1961c), 'Effect of shear and normal forces on the fully plastic moment of an I-beam', *J. Mech. Eng. Sci.*, **3**, 279.

Onat, E. T. and Shield, R. T. (1954), 'The influence of shearing forces on the plastic bending of wide beams', *Proc. 2nd U.S. Nat. Congr. Appl. Mech., Michigan, 1954*, 535.

Prager, W. and Hodge, P. G. (1951), *Theory of Perfectly Plastic Solids*, John Wiley, NY, Chapman & Hall, London, 51.

Rao, N. R. N., Lohrmann, M. and Tall, L. (1966), 'Effect of strain rate on the yield stress of structural steels', *A.S.T.M. J. Materials*, **1**, 241.

Roderick, J. W. and Phillipps, I. H. (1949), 'The carrying capacity of simply supported mild steel beams', *Research (Eng. Struct. Suppl.), Colston Papers*, **2**, 9.

Shakir-Khalil, H. and Tadros, G. S. (1973), 'Plastic resistance of mild steel rectangular sections', *Struct. Engr*, **51**, 239.

Wilson, Carus (1891), 'The influence of surface loading on the flexure of beams', *Phil. Mag.* (ser. 5), **32**, 481.

Examples

1. An I-section has the following dimensions (see Fig. 1.9):

flange width	$B = 178$ mm
flange thickness	$T = 12.8$ mm
total depth	$D = 406$ mm
web thickness	$t = 7.8$ mm.

Find the plastic moment for bending about the minor axis in the absence of axial force and also when there is an axial thrust of 1200 kN, assuming a yield stress of 250 N/mm².

2. A tee section consists of a flange of width a and thickness $0.2a$, and a web of depth a and thickness $0.2a$, the total depth of the section thus being $1.2a$. For bending about an axis parallel to the flange, causing flexure in the plane of the web, determine the plastic moment. Find the interaction relation between plastic moment and axial force, referring the moment to the equal area axis.

3. A thin-walled tube of circular cross section carries an axial force N and a bending moment M_N about a diametral axis. Find the interaction relation between M_N and N.

4. The frame whose dimensions and loading were specified in Fig. 4.7(a) was shown in Section 4.3.4 to collapse at a load factor of 1.342 if the plastic moments of the members had the trial values given in the figure.

If the load factor is to be 1.5, show that the sections of members CE and DF can be a 152×152 @ 23 UC, assuming a lower yield stress of $250 \, \text{N/mm}^2$, when due allowance is made for the axial thrust in these members. For this section, the following properties may be assumed:

$$A = 29.8 \, \text{cm}^2, \quad Z_\text{p} = 184.3 \, (1 - 1.972 \, n^2), \quad n \leqslant 0.283.$$

5. A fixed-ended beam is of length 4.5 m and carries a concentrated load W at a position 1.5 m from one end. The section is a 305×165 @ 54 UB whose relevant properties are:

$$\text{major axis} \quad Z_\text{p} = 843 \, \text{cm}^3$$
$$\text{web area} \quad A_\text{w} = 22 \, \text{cm}^2$$
$$\text{flange area} \quad A_\text{f} = 46 \, \text{cm}^2.$$

Assuming a yield stress of $250 \, \text{N/mm}^2$, find the value of W which would cause plastic collapse (a) neglecting the effect of shear on the plastic moment, and (b) taking shear into account. Assume the von Mises yield criterion and neglect the effect of contact stresses beneath the load.

7 Minimum Weight Design

7.1 Introduction

In the methods of plastic design described in Chapter 4 it was assumed that the characteristic loads acting on a frame were given, and that the problem was to find the plastic moments of the members so that the frame would just collapse if these loads were all multiplied by a specified load factor, say, λ^*. The procedure was to assume trial values for the plastic moments of all the members, and to determine the corresponding load factor λ_c against collapse. The trial values of all the plastic moments were then multiplied by λ^*/λ_c, thereby producing a design in which the load factor against plastic collapse was λ^*.

This procedure involved at the outset the arbitrary selection of the relative values of the plastic moments of the members. A different selection of these relative values would result in a different design Clearly for a given frame and loading there will be a large number of possible designs, and it is relevant to consider which of these represents the best which can be achieved.

The design which involves the use of the least weight of material is one kind of optimum, but to assert that minimum weight is the only important criterion in design would be to disregard the economic and other factors which must always be considered. However, a discussion of those factors will not be entered into here, and this chapter is concerned solely with the problem of designing for minimum structural weight.

7.2 Assumptions

It will be assumed throughout that the dimensions of the frame and the factored values of the characteristic loads are prescribed. Furthermore, each frame considered will be assumed to be composed of uniform prismatic members. The normal assumptions for the calculation of plastic collapse loads are made, and the plastic moments of the members are taken to be unaffected by axial and shear forces.

A relationship between the weight w per unit length of a member and its plastic moment M_p is required. For a series of geometrically similar cross sections, the cross-sectional area and thus w is proportional to d^2, where d is any typical dimension such as the total depth of the section, and the plastic section modulus Z_p and thus M_p is proportional to d^3. It follows that for such a series of sections

$w \propto M_p^{2/3}$. For I-sections it is found that the empirical relation

$$w \propto M_p^{0.6} \tag{7.1}$$

agrees closely with tabulated values.

For a range of sections appropriate to a specific problem, Equation (7.1) may be approximated to by the linear relationship

$$w = a + bM_p \tag{7.2}$$

where a and b are constants. The error involved in this approximation is small, and it will be seen later that its use may lead to the correct minimum weight design.

If it is further assumed that an infinite range of sections is available, the linearization implied by Equation (7.2) permits the formulation of a simple expression for the total weight of a structure. If L is the length of any member, the total structural weight X is given by

$$X = \sum wL = a \sum L + b \sum M_p L,$$

where the summations cover all the members.

The term $a\Sigma L$ is a constant for given structural dimensions, and so X is minimized when $\Sigma M_p L$ is minimized. This term is called the *weight function,* and is denoted by x, so that

$$x = \sum M_p L. \tag{7.3}$$

The minimum weight problem is thus to design a frame so that the weight function x, defined by Equation (7.3), is a minimum.

7.3 Geometrical analogue and Foulkes' theorem

7.3.1 *Geometrical analogue: rectangular frame*

The nature of the minimum weight design problem was clarified by the work of Foulkes (1953, 1954), using a geometrical analogue. His treatment will be followed here with only minor variations. Consider as an illustrative example the rectangular frame whose dimensions and factored loads, in arbitrary units, are shown in Fig. 7.1. The beam is to have a plastic moment β_1, and the columns are each to have a plastic moment β_2. The weight function as defined in Equation (7.3) is

$$x = 3\beta_1 + 2\beta_2. \tag{7.4}$$

For this type of frame and loading there are two independent mechanisms, the beam and sway mechanisms, and these can be combined to form a third mechanism. Since it is not known at the outset whether β_1 is less than or greater than β_2, each of these mechanisms can take two forms, any hinge occurring at a joint between the beam and a column appearing either in the beam or the column. The resulting six possible mechanisms of collapse are shown in Fig. 7.1 together with the corresponding work equations, with θ cancelled on each side.

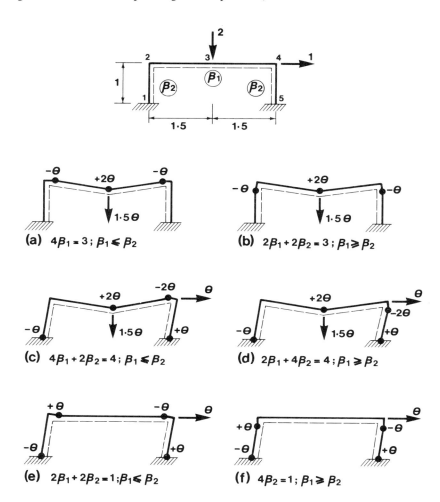

Fig. 7.1 *Rectangular frame: mechanisms and work equations*

These work equations may be represented as straight lines on a diagram in which β_1 and β_2 are rectangular coordinate variables, as in Fig. 7.2. For each of the three types of mechanism there are two lines, representing the work equations for the cases $\beta_1 \leqslant \beta_2$ and $\beta_1 \geqslant \beta_2$. Thus for the sway mechanism the lines (e) and (f) represent these two cases, the lines intersecting at the point N where $\beta_1 = \beta_2$. The other work equations are shown similarly as the lines (a), (b), (c) and (d), the lettering corresponding to that of Fig. 7.1.

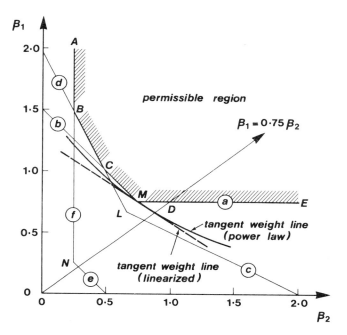

Fig. 7.2 *Geometrical analogue for frame of Fig. 7.1*

A simple deduction can be made from this figure using the kinematic theorem. Consider, for example, a particular case $\beta_1 = 0.75\,\beta_2$. This condition is represented by a straight line through the origin of slope 0.75, which intersects the three work equation lines (e), (c) and (a). The intersection with (a) at the point D is farthest from the origin and thus represents the highest values for the plastic moments. By the kinematic theorem this point represents the required values of β_1 and β_2 if $\beta_1 = 0.75\,\beta_2$. Considering all possible ratios of β_1 to β_2 in this way, it follows that all possible conditions of collapse are represented by the line segments ABCME shown shaded in the figure. Points farther from the origin than these line segments represent designs which would not collapse under the given loads; the region so defined is termed the *permissible region*. Points nearer to the origin than the boundary of the permissible region represent structures which could not support the factored loads.

The minimum weight problem therefore reduces to locating the point on the boundary of the permissible region for which the weight function $x = 3\beta_1 + 2\beta_2$ is a minimum. Any straight line of the form

$$3\beta_1 + 2\beta_2 = \text{constant}$$

represents designs of constant weight, the perpendicular distance from the origin to the line being proportional to the weight function x. Hence the minimum weight design is found by determining where such a line, of slope $-2/3$, just touches the boundary of the permissible region. The appropriate *tangent weight line* is shown broken in Fig. 7.2, touching the boundary of the permissible region at the minimum weight design point M. The coordinates of M are

$$\beta_1 = 0.75, \quad \beta_2 = 0.75$$

and these values of the plastic moments constitute the minimum weight design.

If the relation (7.1) is used to determine the structural weight, lines of constant weight are of the form

$$3\beta_1^{0.6} + 2\beta_2^{0.6} = \text{constant.}$$

The corresponding tangent weight line is shown in Fig. 7.2, and it will be seen that the minimum weight design still corresponds to the point M.

The boundary of the permissible region is convex towards the origin, so that when a straight line of constant weight touches this boundary at one point it is impossible for the line to pass into the permissible region anywhere along its length. This property of convexity is always possessed by the boundary of the permissible region. A design can therefore be established as being of minimum weight by making only local tests to show that the weight is increased by any small changes in the ratios of the plastic moments. However, Prager (1956) has pointed out that this is not always true if a power law weight relationship such as Equation (7.1) is used.

For frames whose design is specified by the values of only two plastic moments, the minimum weight design can always be determined by a graphical procedure of the kind embodied in Fig. 7.2. However, when there are more than two parameters involved, a graphical procedure is impracticable. To enable such cases to be dealt with, and to enable a local test to be used to check a minimum weight design, a theorem due to Foulkes (1954) may be used.

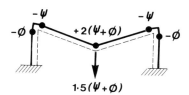

Fig. 7.3 *Mechanism with two degrees of freedom*

7.3.2 *Foulkes' theorem*

Figure 7.2 will be used to introduce this theorem. At the minimum weight point M, the lines corresponding to mechanisms (a) and (b) intersect to form a corner of the boundary of the permissible region. This implies that the minimum weight structure can fail by either of these two mechanisms. These are seen from Fig. 7.1 to be the two possible beam mechanisms, with the hinges at the ends of the beam forming in either the beam, for $\beta_1 \leqslant \beta_2$, or the columns, for $\beta_1 \geqslant \beta_2$. When $\beta_1 = \beta_2$, a collapse mechanism with two degrees of freedom can therefore occur, as shown in Fig. 7.3. This is formed by the addition of the mechanisms of Fig. 7.1(a) and (b), replacing θ by ψ in (a) and by ϕ in (b). The work equation for this mechanism is

$$3(\psi + \phi) = (4\psi + 2\phi)\beta_1 + 2\phi\beta_2;$$
$$\psi \geqslant 0 \quad \text{and} \quad \phi \geqslant 0. \tag{7.5}$$

If $\beta_1 = \beta_2$, this equation gives $\beta_1 = \beta_2 = 0.75$, as obtained previously.

Equation (7.5) is valid within the limits stated, namely $\psi \geqslant 0$ and $\phi \geqslant 0$, these conditions being required to produce hogging moments at each end of the beam. For any given positive ratio of ψ to ϕ it represents a straight line in Fig. 7.2. When $\phi = 0$, the line becomes line (a), corresponding to mechanism (a) of Fig. 7.1; and when $\psi = 0$, it becomes line (b), corresponding to mechanism (b). The slope of the tangent weight line lies between the slopes of these two mechanism lines (a) and (b), and so there must exist a ratio of ψ to ϕ for which Equation (7.5) becomes a line of the same slope as the tangent weight line. From Equation (7.4) the tangent weight line is of the form

$$3\beta_1 + 2\beta_2 = \text{constant.}$$

For Equation (7.5) to represent a line of the same slope it follows by comparing coefficients of β_1 and β_2 that

$$\frac{4\psi + 2\phi}{3} = \frac{2\phi}{2},$$

from which $\phi = 4\psi$. With this value of ϕ, the work equation (7.5) becomes

$$15\psi = (12\beta_1 + 8\beta_2)\psi \tag{7.6}$$

and this represents a line parallel to the lines of constant weight.

The particular beam mechanism obtained by choosing $\phi = 4\psi$ is said to be *weight compatible,* since in its work equation (7.6) the coefficient of each plastic moment bears the same ratio to the corresponding coefficient in the weight function. A general definition of weight compatibility is as follows: If a frame has n different plastic moments $(\beta_1, \beta_2, \ldots, \beta_n)$, the work equation for any mechanism will be of the form

$$\text{Work done} = [c_1\beta_1 + c_2\beta_2 + \ldots + c_n\beta_n]\,\theta,$$

where $c_r\theta$ is the total hinge rotation in all the members whose plastic moment is β_r.

The weight function will be

$$x = [L_1\beta_1 + L_2\beta_2 + \ldots + L_n\beta_n],$$

where L_r is the total length of all the members whose plastic moment is β_r. The mechanism is weight compatible if

$$\frac{c_1}{L_1} = \frac{c_2}{L_2} = \ldots = \frac{c_n}{L_n}.$$

The minimum weight design in this particular case is distinguished by two features. In the first place, there exists a weight compatible mechanism, as has just been seen. Second, the minimum weight design point M lies on the boundary of the permissible region, and this implies that a safe and statically admissible distribution of bending moment throughout the structure can be found. These two features exemplify the theorem due to Foulkes (1954):

Foulkes' theorem. If for any design of a frame a weight-compatible mechanism can be formulated, and a corresponding safe and statically admissible distribution of bending moment throughout the frame can be found, the design will be of minimum weight.

This wording covers the possibility of there being a range of designs all of which have the same minimum weight. Reference may be made to the original paper by Foulkes for other minimum weight theorems, especially those which give upper and lower bounds on the minimum weight.

It would be incorrect to interpret Foulkes' theorem as implying that all minimum weight designs are weight compatible, as emphasised by Smith (1974). The frame of Fig. 7.4, taken directly from Smith's work, illustrates the point. The two columns, of unequal length, are presumed to have different plastic moments, and the beam has a plastic moment equal to that of the shorter column. There are five possible collapse mechanisms, as shown, and their work equations are represented in Fig. 7.5.

The minimum weight design point M in this instance is

$$\beta_1 = 0, \quad \beta_2 = 0.50,$$

and it will be seen that the condition of weight compatibility is not met by the minimum weight design.

7.4 Methods of solution

The problem of design for minimum weight is amenable to the techniques of linear programming. Thus for the frame of Fig. 7.1 the equations of equilibrium are

$$1 = -M_1 + M_2 - M_4 + M_5. \tag{7.7}$$

$$3 = -M_2 + 2M_3 - M_4. \tag{7.8}$$

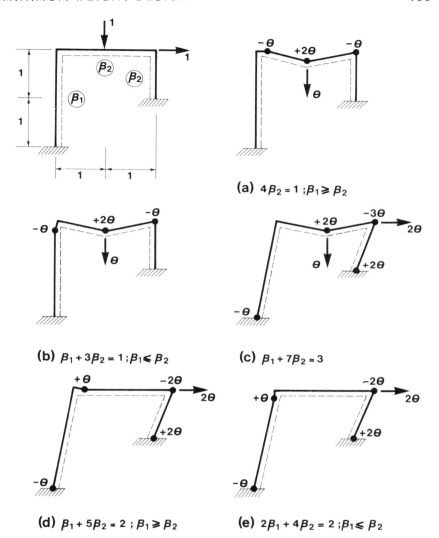

Fig. 7.4 *Frame with columns of different lengths: mechanisms and work equations*

Since no bending moment can exceed the plastic moment in magnitude

$$-\beta_1 \leqslant M_i \leqslant \beta_1 \qquad (i = 2, 3, 4). \tag{7.9}$$

$$-\beta_2 \leqslant M_j \leqslant \beta_2 \qquad (j = 1, 2, 4, 5). \tag{7.10}$$

The problem is therefore to minimize the weight function

$$x = 3\beta_1 + 2\beta_2 \tag{7.4}$$

subject to the constraints imposed by Equations (7.7) and (7.8) and the inequalities (7.9) and (7.10).

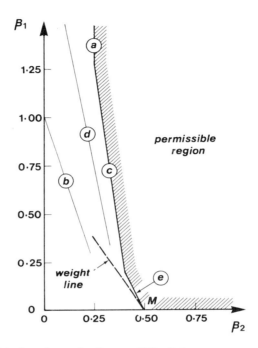

Fig. 7.5 *Geometrical analogue for frame of Fig. 7.4*

This formulation of the problem was used by Heyman (1951, 1952), who gave the first systematic method of solution. Trial and error procedures were suggested by both Foulkes (1953) and Heyman (1953). These depended upon identifying the minimum weight design as satisfying the conditions of Foulkes' theorem, but as pointed out above a minimum weight design is not necessarily weight compatible. It is suggested that recourse should be had to a computer program for any problem beyond the scope of the graphical procedure described in Section 7.3.1.

Livesley (1956) was the first to devise a program for the minimum weight design of frames, and Heyman and Prager (1958) described a technique which can be used manually and which is also suitable for programming on a computer. A comprehensive discussion of the use of linear programming in this context has been given by Maier, Srinivasan and Save (1972), and primal-dual programs have been discussed by Munro (1973).

The minimum weight design problem discussed in this chapter was based on the following assumptions:

(a) the frame is composed of uniform prismatic members;

(b) an infinite range of sections is available;

(c) plastic collapse should occur under a single specified combination of factored loads;

(d) the linearized weight function is to be minimized.

Many studies have been made of minimum weight problems with different formulations. For example, if assumption (a) is not made, so that varying sections may be used, further reductions in weight are possible. Horne (1952) considered the problem of a fixed-ended beam with additional flange plates (see Section 3.5.4). The use of tapered members was investigated by Vickery (1962), and Heyman (1959, 1960) has considered the absolute minimum weight which would result from the use of continuously varying sections.

Toakley (1968) discussed the problem when assumption (b) is discarded, and due account is taken of the fact that in practice only a finite range of sections is available. Another variant of practical significance occurs when assumption (c) is altered to cover the case of a frame which must be capable of withstanding a single application of two or more combinations of factored loads. This problem was considered by Heyman (1951, 1952), Foulkes (1955) and also by Livesley (1959). It should not be confused with the case of variable repeated loading which is dealt with in Chapter 8.

References

Foulkes, J. (1953), 'Minimum weight design and the theory of plastic collapse', *Q. Appl. Math.*, **10**, 347.

Foulkes, J. (1954), 'The minimum weight design of structural frames', *Proc. R. Soc.*, A., **223**, 482.

Foulkes, J. (1955), 'Linear programming and structural design', *Proc. 2nd Symp. Linear Programming, Washington, 1955,* 177.

Heyman, J. (1951), 'Plastic design of beams and plane frames for minimum material consumption', *Q. Appl. Math.*, **8**, 373.

Heyman, J. (1952), 'Plastic analysis and design of steel-framed structures', prelim. publ., 4th Congr. Int. Assoc. Bridge Struct. Eng., Cambridge, 1952, 95.

Heyman, J. (1953), 'Plastic design of plane frames for minimum weight', *Struct. Engr*, **31**, 125.

Heyman, J. (1959), 'On the absolute minimum-weight design of framed structures', *Q. J. Mech. Appl. Math.*, **12**, 314.

Heyman, J. (1960), 'On the minimum-weight design of a simple portal frame', *Int. J. Mech. Sci.*, **1**, 121.

Heyman, J. and Prager, W. (1958), 'Automatic minimum-weight design of steel frames', *J. Franklin Inst.*, **266**, 339.

Horne, M. R. (1952), 'Determination of the shape of fixed-ended beams for maximum economy according to the plastic theory', prelim publ., 4th Congr. Int. Assoc. Bridge Struct. Eng., Cambridge, (1952), 111 (see also Final Rep., 119, 1952).

Livesley, R. K. (1956), 'The automatic design of structural frames', *Q. J. Mech. Appl. Math.*, **9**, 257.

Livesley, R. K. (1959), 'Optimum design of structural frames for alternative systems of loading', *Civ. Eng. Publ. Wks. Rev.*, **54**, 737.

Maier, G., Srinivasan, R. and Save, M. A. (1972), 'On limit design of frames using linear programming', Proc. Int. Symp. Computer-aided Structural Design, vol. 1, Warwick Univ., 1972.

Munro, J. (1973), 'The analysis and synthesis of safe and serviceable structures', Proc. NATO Adv. Study Inst. Generic Techniques in Systems Reliability Assessment, Liverpool, 1973.

Prager, W. (1956), 'Minimun-weight design of a portal frame', *J. Eng. Mech. Div., Proc. Am. Soc. Civil Engrs*, Paper 1073.

Smith, D. L. (1974), 'Plastic limit analysis and synthesis of structures by linear programming', Ph.D. thesis, London Univ.

Toakley, A. R. (1968), 'Optimum design using available sections', *J. Struct. Div., Proc. Am. Soc. Civil Engrs*, **94** (ST5), 1219.

Vickery, B. J. (1962), 'The behaviour at collapse of simple steel frames with tapered members', *Struct. Engr*, **40**, 365.

Examples

1. A beam ABC rests on three simple supports, A, B and C, where AB = BC = 3 m. Concentrated vertical loads 60 kN and 50 kN are applied 1 m from A and 1.5 m from C, respectively. If the plastic moments of the spans AB and BC are β_1 and β_2, respectively, find the values of β_1 and β_2 in the minimum weight design.

2. A fixed-base rectangular frame ABCD has columns AB and DC of equal height 4 m and a horizontal beam BC of length 8 m. There is a concentrated vertical load 40 kN at the centre of the beam and a concentrated horizontal load 30 kN at C in the direction BC. If the plastic moment of the beam is β_1, and the plastic moments of the columns are each to be equal to β_2, find the minimum weight design.

3. If the frame of example 2 is freely pinned to a rigid base at D, while all other conditions remain unchanged, find the values of β_1 and β_2 in the minimum weight design.

4. If in the frame of example 2 the columns AB and DC may have different plastic moments β_2 and β_3, respectively, while the beam BC has a plastic moment β_1, show that the minimum weight design is $\beta_1 = \beta_3 = 55, \beta_2 = 5$.

5. The frame of Fig. 4.1(a) was analysed in Section 4.2 on the assumption that all the members had the same plastic moment. Show that if the two rafters are required to have the same plastic moment β_1, and the two columns are required to have the same plastic moment β_2, the minimum weight design is achieved when $\beta_1 = \beta_2$.

6. In the frame of Fig. 4.7(a), the plastic moments of the members are to be:

$$CA, AB, BD: \beta_1$$

$$CE, DF \qquad 2\beta_1$$

$$CD \qquad \beta_2.$$

This frame was analysed in Section 4.3.4 assuming that $\beta_2 = 3\beta_1$. Show that this analysis leads to the minimum weight design.

7. A beam ABCD rests on four simple supports A, B, C and D, where AB $= 3$ m, BC $= 4$ m, CD $= 5$ m. Concentrated vertical loads 40, 30 and 30 kN are applied at the centres of the spans AB, BC and CD, respectively, and the corresponding plastic moments in these spans are β_1, β_2 and β_3. Show that there is a range of designs all of which are of minimum weight, given by $15 \leqslant \beta_2 \leqslant 20$, $\beta_1 = 30 - 0.5\beta_2$, $\beta_3 = 37.5 - 0.5\beta_2$.

8 Variable Repeated Loading

8.1 Introduction

The loading on a structure may vary considerably during its lifetime. For example, apart from dead-loading a building frame will experience snow loads on the roof and wind loads on each face. The magnitudes of these loads at any particular instant cannot be foreseen, although their characteristic values will be known, so that the sequence of loading is unpredictable. This type of loading is termed *variable repeated loading.*

It is possible, as first recognized by Grüning (1926) and Kazinczy (1931), that under variable repeated loading a frame may fail due to the eventual development of excessive plastic flow, even though no single load combination is sufficiently severe to cause failure by plastic collapse. To aid discussion, suppose that a frame is subjected to loads λP_1, λP_2, . . . , λP_r, . . . , λP_n, each load being applied at a given point in a specified direction. λ is a load factor applicable to every load. The value of any load λP_r can vary between limits $(\lambda P_r^{\max}, \lambda P_r^{\min})$, independently of the variations which may occur in the values of the other loads. The limits $(P_r^{\max}, P_r^{\min})$ are presumed to be prescribed (characteristic) values.

There are two ways in which failure can occur due to variable repeated loading. If the loads on a frame are alternating in character, one or more members may be bent back and forth repeatedly, so that yield occurs in its fibres alternately in tension and compression. This behaviour, termed *alternating plasticity,* may eventually lead to failure by low endurance fatigue. There will be an alternating plasticity load factor λ_a above which alternating plasticity will occur.

Another type of failure may occur if critical combinations of loads follow one another in fairly definite cycles. If λ exceeds a certain value λ^*, increments of plastic hinge rotation take place at certain cross sections during each cycle of loading. These increments are in the same sense in every cycle. If λ, while exceeding λ^*, is less than a higher critical value λ_I, the increments of rotation become progressively smaller as the number of cycles increases. Eventually, a condition is reached in which there are no further changes in the hinge rotations, and during subsequent cycles there are only elastic changes of bending moment in the frame. When this happens, the frame is said to have *shaken down.* However, if λ exceeds λ_I, the frame never shakes down, and finite rotations occur at the hinges during each cycle. Thus if a sufficient number of cycles of loading occurs, unacceptably large hinge rotations and therefore deflections will be built up. The frame would then be said to have failed by *incremental collapse*. The critical load factor λ_I above which incremental collapse can occur is termed the *incremental collapse load factor*.

The main objective of this chapter is to discuss methods for the determination of λ_I, since the calculation of λ_a is found to be a simple matter. However, an essential preliminary is to develop an understanding of the behaviour of frames under cyclic loading when λ exceeds λ^*. Section 8.2 is therefore devoted to a description of the response of a particular frame to this type of loading. This is followed by a discussion in Section 8.3 of the theorems concerning the values of λ_I and λ_a. It appears that a close parallel exists between these theorems and those which concern the collapse load factor λ_c for a particular load combination. It follows that methods for the calculation of λ_I can be developed which are analogous to those given in Chapter 4 for the calculation of λ_c. One such method is described in Section 8.4. In conclusion, a discussion is given in Section 8.5 of the significance of the phenomena of alternating plasticity and incremental collapse in relation to plastic design.

8.2 Step-by-step calculations

The results of some illustrative step-by-step calculations for the fixed-base rectangular frame shown in Fig. 8.1 will now be presented. Each member of this frame is presumed to have the same plastic moment $\pm M_p$ and to behave elastically, with flexural rigidity EI, when the magnitude of the bending moment is less than M_p. This frame was analysed in Section 2.5 under proportional loading. The same step-by-step technique is applicable for the analysis of cyclic loading, and so details of the calculations need not be given.

The bending moments which occur if the response of the frame is wholly

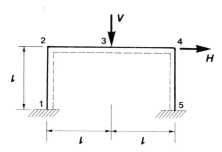

Fig. 8.1 *Frame and loading*

elastic, which are denoted by the symbol $\mathcal{M}$, will be needed in what follows. They may be shown to be:

$$\mathcal{M}_1 = +0.1\,Vl - 0.3125\,Hl$$
$$\mathcal{M}_2 = -0.2\,Vl + 0.1875\,Hl$$
$$\mathcal{M}_3 = +0.3\,Vl \qquad\qquad\qquad (8.1)$$
$$\mathcal{M}_4 = -0.2\,Vl - 0.1875\,Hl$$
$$\mathcal{M}_5 = +0.1\,Vl + 0.3125\,Hl.$$

8.2.1 *Alternating plasticity*

The first loading cycle to be considered is shown in Fig. 8.2. The sequence of loading is:

$$V = W, H = W \quad \text{or simply} \quad (W, W)$$
$$V = 0, \ H = 0 \qquad\qquad\quad (0, 0)$$
$$V = 0, \ H = -W \qquad\quad (0, -W)$$
$$V = 0, \ H = 0 \qquad\qquad\quad (0, 0).$$

This cycle causes alternating plasticity if W exceeds a critical value W_a.

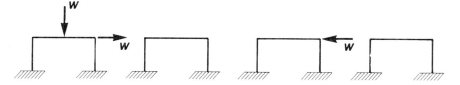

Fig. 8.2 *Cycle of loading which may cause alternating plasticity*

The results of calculations for $W = 2.85\,M_p/l$ are summarized in Table 8.1. When the load combination (W, W) is first applied,

Table 8.1 *Alternating plasticity: $W = 2.85\,M_p/l$*

$\dfrac{Vl}{M_p}$	$\dfrac{Hl}{M_p}$	$\dfrac{M_1}{M_p}$	$\dfrac{M_2}{M_p}$	$\dfrac{M_3}{M_p}$	$\dfrac{M_4}{M_p}$	$\dfrac{M_5}{M_p}$	$\dfrac{\phi_4 EI}{M_p l}$	$\dfrac{\phi_5 EI}{M_p l}$
2.85	2.85	−0.823	0.028	0.939	−1	1	−0.158	0.103
0	0	−0.217	0.063	0.084	0.104	−0.176	−0.158	0.103
0	−2.85	0.715	−0.491	0.077	0.645	−1	−0.158	0.069
0	0	−0.176	0.044	0.077	0.110	−0.109	−0.158	0.069
2.85	2.85	−0.823	0.028	0.939	−1	1	−0.158	0.103

hinges form and rotate at sections 5 and 4, their signs being positive and negative, respectively. This is followed by elastic unloading. During the application of the load combination $(0, -W)$ the bending moment at section 5 reaches the value $-M_p$, and there is a consequent change of $-0.034 M_p l/EI$ in the hinge rotation at this section. The bending moment at section 4 remains within the elastic range, and so the hinge rotation at this section is unaltered.

Elastic unloading completes the first cycle. During the second application of the loading (W, W), the bending moment at section 5 reaches the value $+M_p$, and there is a change of $+0.034 M_p l/EI$ in the hinge rotation at this section. The bending moment at section 4 only just attains the value $-M_p$ as the peak loads are reached, and so there is no change in the corresponding hinge rotation. The resulting bending moments and hinge rotations are given in the fifth row of Table 8.1, and it will be seen that they are identical with those occurring after the first application of the loading (W, W). A condition of alternating plasticity has therefore been established, in which the plastic hinge rotation at section 5 varies between the extreme values $0.103 M_p l/EI$ and $0.069 M_p l/EI$ during each cycle of loading.

In this particular case the critical value W_a of W, above which alternating plasticity would occur at section 5, is readily calculated. If W was equal to W_a, then the change of loading from (W, W) to $(0, -W)$ would just change the bending moment at section 5 from M_p to $-M_p$, with the entire frame behaving elastically during this load change. Using superscripts to represent the load combinations, it is seen from Equations (8.1) that

$$\mathcal{M}_5^{(W, W)} = 0.4125\, Wl, \quad \mathcal{M}_5^{(0, -W)} = -0.3125\, Wl.$$

Thus when $W = W_a$,

$$(0.4125\, W_a l + 0.3125\, W_a l) = 2M_p$$

$$W_a = 2.759\, M_p/l.$$

8.2.2 Incremental collapse

A loading cycle which may cause incremental collapse if W exceeds a critical value W_I is shown in Fig. 8.3. The sequence of loading is:

$$V = W, H = W \quad \text{or} \quad (W, W)$$

$$V = 0, H = 0 \quad\quad (0, 0)$$

$$V = 0, H = W \quad\quad (0, W)$$

$$V = 0, H = 0 \quad\quad (0, 0).$$

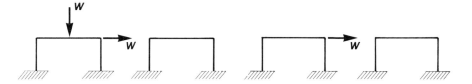

Fig. 8.3 *Cycle of loading which may cause incremental collapse*

The results of calculations for $W = 2.9M_p/l$ are summarized in Table 8.2. In each cycle of loading, application of the load combination (W, W) causes the formation and rotation of plastic hinges at sections 4 and 5. Additionally, from the third cycle onwards, this loading also causes a plastic hinge to form and rotate at section 3. After elastic unloading, the loading $(0, W)$ causes the formation and rotation of a plastic hinge at section 1.

The changes of bending moment during the third and fourth cycles are identical, and recur in further cycles of the same loading. The changes in hinge rotation, denoted by $\Delta\phi$, which occur during the fourth cycle are as follows:

section	$\Delta\phi EI/M_p l$	loading
1	−0.045	$(0, W)$
3	+0.090	(W, W)
4	−0.090	(W, W)
5	+0.045	(W, W)

The same changes in hinge rotation would occur in each subsequent cycle of the same loading. Putting $\alpha = 0.045\,M_p l/EI$, these changes are as depicted in Fig. 8.4(a). It will be seen that if they took place simultaneously, they would constitute a mechanism motion.

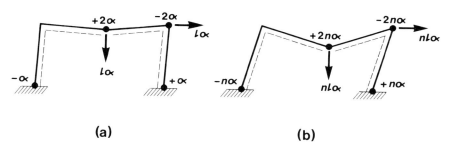

(a)   (b)

Fig. 8.4 *Incremental collapse*
    (a) Changes during one cycle
    (b) Changes during n cycles

Table 8.2 Incremental collapse: $W = 2.9 M_p/l$

$\dfrac{Vl}{M_p}$	$\dfrac{Hl}{M_p}$	$\dfrac{M_1}{M_p}$	$\dfrac{M_2}{M_p}$	$\dfrac{M_3}{M_p}$	$\dfrac{M_4}{M_p}$	$\dfrac{M_5}{M_p}$	$\dfrac{\phi_1 EI}{M_p l}$	$\dfrac{\phi_3 EI}{M_p l}$	$\dfrac{\phi_4 EI}{M_p l}$	$\dfrac{\phi_5 EI}{M_p l}$
2.9	2.9	−0.865	0.035	0.968	−1	1	0	0	−0.186	0.116
0	0	−0.249	0.071	0.098	0.124	−0.196	0	0	−0.186	0.116
0	2.9	−1	0.629	0.082	−0.465	0.806	−0.078	0	−0.186	0.116
0	0	−0.094	0.085	0.082	0.079	−0.100	−0.078	0	−0.186	0.116
2.9	2.9	−0.818	0.082	0.991	−1	1	−0.078	0	−0.256	0.171
0	0	−0.202	0.118	0.121	0.124	−0.196	−0.078	0	−0.256	0.171
0	2.9	−1	0.672	0.110	−0.451	0.777	−0.133	0	−0.256	0.171
0	0	−0.094	0.128	0.110	0.092	−0.129	−0.133	0	−0.256	0.171
2.9	2.9	−0.800	0.100	1	−1	1	−0.133	0.050	−0.333	0.216
0	0	−0.184	0.136	0.130	0.124	−0.196	−0.133	0.050	−0.333	0.216
0	2.9	−1	0.688	0.121	−0.446	0.766	−0.179	0.050	−0.333	0.216
0	0	−0.094	0.144	0.121	0.098	−0.140	−0.179	0.050	−0.333	0.216
2.9	2.9	−0.800	0.100	1	−1	1	−0.179	0.140	−0.423	0.261
0	0	−0.184	0.136	0.130	0.124	−0.196	−0.179	0.140	−0.423	0.261
0	2.9	−1	0.688	0.121	−0.446	0.766	−0.224	0.140	−0.423	0.261
0	0	−0.094	0.144	0.121	0.098	−0.140	−0.224	0.140	−0.423	0.261

The horizontal deflection h at the top of each column is therefore increased by an amount $l\alpha$ in each complete cycle of loading after the third. The effect of n such cycles would therefore be to increase h by

$$n l\alpha = n \left[\frac{0.045\,M_\mathrm{p}l^2}{EI} \right],$$

as shown in Fig. 8.4(b). It follows that if n is sufficiently large, deflections of any magnitude can be built up, with the same increment occurring in each cycle. This would be a failure by incremental collapse, and the mechanism of Fig. 8.4(b) is therefore referred to as the incremental collapse mechanism.

Since the eventual deflections of the frame are very large, the requirements of compatibility can only be met because the increments of plastic hinge rotation which take place in each cycle are consistent with a mechanism motion, which of itself demands no curvature changes along the members. This is the essential characteristic of any failure by incremental collapse.

The phenomenon of incremental collapse may best be understood by examining the changes in residual moments which occur during a cycle of loading. Consider the third cycle in Table 8.2. When the loads (W, W) are removed the entire frame behaves elastically, and at section 5, for example, the bending moment changes from M_p to $-0.196\,M_\mathrm{p}$, an elastic change of $-1.196\,M_\mathrm{p}$. If this same loading were to be reapplied, the bending moment at section 5 would just reach the value M_p as the peak loads were attained, since the elastic changes are reversible. Using the symbol m to denote residual moment (under zero external load), the situation just described is summarized as follows:

$$m_5 = -0.196 M_\mathrm{p}$$
$$\mathscr{M}_5^{(W, W)} = +1.196 M_\mathrm{p} \quad (\text{for } W = 2.9\,M_\mathrm{p}/l)$$
$$m_5 + \mathscr{M}_5^{(W, W)} = -0.196 M_\mathrm{p} + 1.196 M_\mathrm{p} = M_\mathrm{p}.$$

However, in the cycle of loading under consideration, removal of the loads (W, W) is followed by the loading $(0, W)$. This causes the formation and rotation of a plastic hinge at section 1, and the residual moment distribution throughout the frame is thereby altered. In particular, m_5 becomes $-0.140 M_\mathrm{p}$. Thus when the loading (W, W) is subsequently applied, the frame cannot respond by wholly elastic behaviour, since as far as section 5 is concerned this would require

$$m_5 + \mathscr{M}_5^{(W, W)} \leqslant M_\mathrm{p},$$

whereas in fact

$$m_5 + \mathscr{M}_5^{(W, W)} = -0.140 M_\mathrm{p} + 1.196 M_\mathrm{p} = 1.056 M_\mathrm{p}.$$

Further examination of Table 8.2 shows that whenever the loading (W, W) is applied, the plastic hinge rotations which occur at sections 3, 4 and 5 alter the residual moment at section 1 in such a way that the loading $(0, W)$ causes a

Table 8.3 *Behaviour when* $W = W_I = 2.857 M_p/l$

$\dfrac{Vl}{M_p}$	$\dfrac{Hl}{M_p}$	$\dfrac{M_1}{M_p}$	$\dfrac{M_2}{M_p}$	$\dfrac{M_3}{M_p}$	$\dfrac{M_4}{M_p}$	$\dfrac{M_5}{M_p}$	$\dfrac{\phi_1 EI}{M_p l}$	$\dfrac{\phi_4 EI}{M_p l}$	$\dfrac{\phi_5 EI}{M_p l}$
2.857	2.857	−0.828	0.029	0.943	−1	+1	0	−0.162	+0.105
0	0	−0.221	0.065	0.086	0.107	−0.179	0	−0.162	+0.105
0	2.857	−1	0.610	0.074	−0.462	0.785	−0.058	−0.162	+0.105
0	0	−0.107	0.074	0.074	0.074	−0.108	−0.058	−0.162	+0.105
2.857	2.857	−0.794	0.063	0.960	−1	+1	−0.058	−0.214	+0.145
0	0	−0.187	0.099	0.103	0.107	−0.179	−0.058	−0.214	+0.145
0	2.857	−1	0.642	0.095	−0.452	0.764	−0.098	−0.214	+0.145
0	0	−0.107	0.106	0.095	0.084	−0.129	−0.098	−0.214	+0.145
2.857	2.857	−0.770	0.087	0.972	−1	+1	−0.098	−0.250	+0.173
0	0	−0.163	0.123	0.115	0.107	−0.179	−0.098	−0.250	+0.173
0	2.857	−1	0.664	0.110	−0.445	0.749	−0.126	−0.250	+0.173
0	0	−0.107	0.128	0.110	0.091	−0.144	−0.126	−0.250	+0.173
2.857	2.857	−0.753	0.104	0.981	−1	+1	−0.126	−0.276	+0.193
0	0	−0.146	0.140	0.124	0.107	−0.179	−0.126	−0.276	+0.193
0	2.857	−1	0.679	0.120	−0.440	0.738	−0.146	−0.276	+0.193
0	0	−0.107	0.143	0.120	0.096	−0.155	−0.146	−0.276	+0.193

further plastic hinge rotation at this section. Similarly, whenever the loading $(0, W)$ is applied, the plastic hinge rotation at section 1 alters the residual moments at sections 3, 4 and 5 so that the loading (W, W) causes further plastic hinge rotations at each of these three sections. It is therefore the changes in residual moments which occur as each load combination is applied which lead to incremental collapse for the particular value of W under consideration.

8.2.3 Behaviour when $W = W_I$

The same loading cycle will now be considered, but with a lower value of W, namely $2.857 M_p/l$. It will be shown that this is the limiting value of W above which incremental collapse will occur, and it is therefore termed the *incremental collapse load*.

Table 8.3 summarizes the behaviour during the first four cycles. After the first application of the loading (W, W), the value of M_3 is $0.943 M_p$, and with each successive application of this loading M_3 increases. Table 8.4 gives the value of M_3 as a function of n, the number of applications of the load combination (W, W), and it will be seen that M_3 approaches M_p asymptotically, only reaching M_p after an infinite number of cycles. This behaviour may be contrasted with the response to the same cycle of loading, but with $W = 2.9 M_p/l$, which was given in Table 8.2. With this value of W, M_3 attained the value M_p during the third application of the loading (W, W), and in that and each subsequent cycle plastic hinge rotation occurred at section 3.

Table 8.4 *Values of M_3 for load combination (W, W): $W = 2.857 M_p/l$*

n	1	2	3	4	5	6	7	8	9	10	∞
M_3/M_p	0.943	0.960	0.972	0.981	0.987	0.991	0.994	0.996	0.997	0.998	1

It therefore appears that for this cycle of loading, a plastic hinge will only form at section 3, in addition to sections 1, 4 and 5, when W exceeds $2.857 M_p/l$. Incremental collapse only becomes possible when the effect of a complete cycle of loading is to add increments of plastic hinge rotation at the four sections 1, 3, 4 and 5, these increments corresponding to a mechanism motion if they all occurred simultaneously. It follows that the incremental collapse load W_I is $2.857 M_p/l$.

A further feature of the behaviour summarized in part in Table 8.3 is that the increments in plastic hinge rotation occurring in each cycle decrease in a geometrical progression, and therefore tend to zero as the number of cycles tends to infinity. Thus after an infinite number of cycles have taken place, there is no further plastic flow anywhere in the frame, which would respond elastically to any further cycles of loading. When this happens, the frame is said to have *shaken down*.

Consider now the frame in its shaken down condition, when $W = W_I$. The residual moments which have been attained must satisfy the following conditions:

$$m_1 + \mathcal{M}_1^{(0, W)} = -M_p$$
$$m_3 + \mathcal{M}_3^{(W, W)} = +M_p$$
$$m_4 + \mathcal{M}_4^{(W, W)} = -M_p \qquad (8.2)$$
$$m_5 + \mathcal{M}_5^{(W, W)} = +M_p.$$

Substituting the values of the elastic moments derived from Equations (8.1),

$$m_1 = \quad 0.3125\, W_I l - M_p$$
$$m_3 = -0.3 W_I l + M_p$$
$$m_4 = \quad 0.3875 W_I l - M_p \qquad (8.3)$$
$$m_5 = -0.4125 W_I l + M_p.$$

The residual moments must be statically admissible with zero external loads. There are two equations of equilibrium, which may be obtained in the usual way by the principle of virtual displacements; they are:

$$-m_1 + m_2 - m_4 + m_5 = 0. \qquad (8.4)$$
$$-m_2 + 2m_3 - m_4 = 0. \qquad (8.5)$$

Adding these equations,

$$-m_1 + 2m_3 - 2m_4 + m_5 = 0. \qquad (8.6)$$

Substituting the residual moments given in Equations (8.3), it follows that

$$-(0.3125\, W_I l - M_p) + 2(-0.3\, W_I l + M_p) - 2(0.3875\, W_I l - M_p)$$
$$+ (-0.4125\, W_I l + M_p) = 0$$
$$2.1\, W_I l = 6 M_p$$
$$W_I = 2.857 M_p/l.$$

With this value of W_I established, the residual moments are found from Equations (8.3) and either (8.4) or (8.5) to be:

$$m_1 = -0.107\, M_p$$
$$m_2 = +0.179\, M_p$$
$$m_3 = +0.143\, M_p$$
$$m_4 = +0.107\, M_p$$
$$m_5 = -0.179\, M_p.$$

If the step-by-step analysis, summarized in Table 8.3 for the first four cycles of loading, is continued until the shaken down condition is established, the same residual moments are found to have been developed.

The process just described affords a method of calculation for the value of W_I, which is however dependent upon the knowledge that the incremental collapse mechanism is as shown in Fig. 8.4(b). Theorems which enable the correct incremental collapse mechanism to be identified are discussed in Section 8.3.

8.2.4 *Effect of cyclic loading on deflection*

Once the plastic hinge rotations have been calculated, the deflection at any point is readily determined, for example by the unit load method (Section 2.5.5). Fig. 8.5 shows how the horizontal deflection h which occurs at the top of either column under the loading (W, W) varies with n, the number of applications of this load combination.

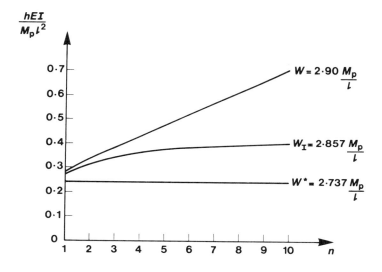

Fig. 8.5 *Effect of cyclic loading on horizontal deflection*

Three (h, n) relations are shown. When $W = 2.9\,M_p/l$, the same increment of deflection occurs in each loading cycle after the third, when the incremental collapse mechanism becomes established. For $W = W_I = 2.857\,M_p/l$, the deflection increases with each cycle, but tends asymptotically to a definite limit as the frame shakes down. Further step-by-step analyses show that the behaviour is similar for values of W lying between the limits W^* and W_I, where $W^* = 2.737 M_p/l$. When W is less than W^*, repetition of the loading cycle produces no increase in deflection. Plastic hinges may form when the load combination (W, W) is first applied, but the residual moments which are then produced are such that all further loadings are borne by wholly elastic action. Calculations of this kind were first given by Horne (1949) for the case of a fixed-ended beam of length l subjected to a load applied alternately at points at a distance $l/3$ from either end of the beam.

It is readily verified that plastic collapse would occur under a single application of the load combination (W, W) if $W = W_c = 3M_p/l$. Thus for this frame and loading,

$$\frac{W_I}{W_c} = \frac{2.857}{3} = 0.95,$$

so that incremental collapse cannot occur unless W is within 5 per cent of the plastic collapse load.

Fig. 8.5 illustrates the general point that the number of cycles of loading required to produce considerable deflections during incremental collapse is quite small. For instance, when $W = 2.90 M_p/l$, the deflection occurring upon the first application of the loading (W, W) is doubled after only seven cycles of loading.

8.2.5 *Experimental evidence*

Neal and Symonds (1958) tested miniature rectangular frames subjected to cycles of loading of the type shown in Fig. 8.3. The results obtained, relating the growth of deflection to the number of cycles of loading at various values of W, agreed well with theoretical relations of the type shown in Fig. 8.5. In particular, it was observed that the growth of deflection was limited until a value of W close to the calculated value of W_I was reached.

Tests on beams resting on three supports, with loads applied to each span, were carried out by Massonet (1953), and also by Gozum and Haaijer (1955). Popov and McCarthy (1960) tested a pinned-base rectangular frame with unequal column lengths. In all these investigations reasonable agreement was found.

8.3 Shake-down theorems

It was shown in Section 8.2 that when a structure is subjected to variable repeated loading, it is possible for plastic flow to continue indefinitely by either alternating plasticity or incremental collapse. The shake-down theorems are concerned with those conditions under which plastic flow will eventually cease, no matter how often and in what sequence the loads are applied. Some definitions are needed before the theorems are stated.

8.3.1 *Definitions*

At any stage during the loading, let M_j denote the bending moment at a particular cross section j. If all the loads were completely removed, and the frame behaved elastically during this removal, there would be a residual bending moment m_j at this section defined by

$$m_j = M_j - \mathcal{M}_j, \tag{8.7}$$

where $\mathcal{M}_j$ is the bending moment which would be produced by the current loading if the entire frame behaved elastically.

It is possible that the unloading process might not be wholly elastic, but even if this were so Equation (8.7) is still taken as the formal definition of residual moment for the present purpose. Any distribution of residual moment m_j defined in this way will be statically admissible with zero external loading, since both M_j and $\mathcal{M}_j$ must be statically admissible with the current loading.

As in Section 8.1, it is supposed that a frame is subjected to loads λP_1, $\lambda P_2, \ldots, \lambda P_r, \ldots, \lambda P_n$, and that each load λP_r can vary between limits ($\lambda P_r^{\max}$, $\lambda P_r^{\min}$). The Principle of Superposition may be used to determine the maximum and minimum possible values of the elastic moments $\mathcal{M}_j$ when all possible variations of the loads between their prescribed limits are taken into account. These values will be denoted by $\lambda.\mathcal{M}_j^{\max}$ and $\lambda.\mathcal{M}_j^{\min}$. The calculation is easily performed when each load λP_r can vary independently of the others, as will be seen in Section 8.4. It is also possible to allow for any connection between two or more of the loads, for example that two loads could never be applied simultaneously.

There will be a shake-down load factor λ_s above which the frame would fail to shake down, with plastic flow continuing either by alternating plasticity or by incremental collapse. Thus:

$$\text{for alternating plasticity } \lambda_s = \lambda_a$$

$$\text{for incremental collapse } \lambda_s = \lambda_I,$$

where λ_a and λ_I are the alternating plasticity and incremental collapse load factors as defined in Section 8.1.

Examples of failures by alternating plasticity and incremental collapse were given in Section 8.2. In both cases these involved the repetition of particular cycles of loading at a constant load factor. Fortunately, the values of λ_I, λ_a and therefore λ_s are independent of the detailed nature of the sequence of loading to which the frame may be subjected, as will be evident from the form of the shake-down theorems.

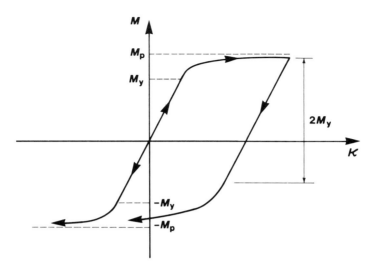

Fig. 8.6 *Bending moment-curvature relation assumed for shake-down theorem.*

The bending moment-curvature relation which is usually assumed in shake-down analysis is shown in Fig. 8.6. This relation is appropriate for a beam of ideal plastic material whose cross section has two axes of symmetry, and which is bent about one of these axes. The magnitudes M_y and M_p of the yield and plastic moments are then the same for bending in either sense. Furthermore, the yield range of bending moment within which wholly elastic behaviour will occur remains at $2M_y$ regardless of the previous loading history.

8.3.2 *Shake-down or lower bound theorem*

For the (M, κ) relation of Fig. 8.6, the following theorem has been established:

Shake-down theorem. If there exists any distribution of residual bending moment $\bar{m}$ throughout a frame which is statically admissible with zero external loading and which also satisfies at every cross section j the conditions

$$\bar{m}_j + \lambda \mathcal{M}_j^{\max} \leqslant (M_p)_j \qquad (8.8)$$

$$\bar{m}_j + \lambda \mathcal{M}_j^{\min} \geqslant -(M_p)_j \qquad (8.9)$$

$$\lambda(\mathcal{M}_j^{\max} - \mathcal{M}_j^{\min}) \leqslant 2(M_y)_j, \qquad (8.10)$$

the value of λ will be less than or equal to the shake-down load factor λ_s.

The conditions (8.8), (8.9) and (8.10) will be referred to collectively as *statical conditions*. It is evident that if they could not be satisfied, shake-down could not occur; these conditions are therefore necessary for shake-down to be possible.

The theorem states that they are also sufficient. There is thus a close analogy with the static theorem of plastic collapse, given in Section 3.2.1. A frame can be said to adapt itself to variable repeated loading in the most effective way possible; plastic flow only continues indefinitely when λ increases beyond the value λ_s above which it becomes impossible to find any set of residual moments satisfying the necessary conditions for the avoidance of plastic flow. Similarly under proportional loading a frame only fails by plastic collapse when the load factor λ increases to a value λ_c above which no safe and statically admissible bending moment distribution can be found. The existence of at least one such distribution is clearly a necessary condition for it to be possible for the loading in question to be carried.

If λ is imagined to be increased steadily from zero, it will become progressively more difficult to satisfy the inequalities (8.8)–(8.10). One possibility is that the inequality (8.10) could not be satisfied at a particular cross section if λ exceeded λ_s, although all the conditions (8.8) and (8.9) could still be met. In this case the failure would be by alternating plasticity, with $\lambda_s = \lambda_a$. The other possibility is that the conditions (8.8) and (8.9) could not all be met simultaneously, although each inequality (8.10) was still satisfied. The failure would then be by incremental collapse, with $\lambda_s = \lambda_I$.

A continued inequality can be written down for each cross section j from the inequalities (8.8) and (8.9), as follows:

$$-(M_p)_j - \lambda \mathcal{M}_j^{\min} \leqslant \bar{m}_j \leqslant (M_p)_j - \lambda \mathcal{M}_j^{\max},$$

so that

$$\lambda(\mathcal{M}_j^{\max} - \mathcal{M}_j^{\min}) \leqslant 2(M_p)_j. \tag{8.11}$$

This may be compared with the inequality (8.10); it is less restrictive but becomes identical with (8.10) for the case of a beam with a shape factor ν of unity, so that $M_y = M_p$. In this special case the inequality (8.10) may be dropped from the shake-down conditions, as it is contained within (8.8) and (8.9).

The shake-down theorem was first stated by Bleich (1932), but his proof only covered frames with not more than two redundancies. A general proof for hypothetical pin-jointed trusses, assuming ideal plastic member behaviour in both tension and compression, was given by Melan (1936), and was later simplified by Symonds and Prager (1950). Melan's proof was adapted by Neal (1950, 1951) to frames, and Appendix B sets out the details for the special case $\nu = 1$, so that $M_y = M_p$.

As pointed out by Koiter (1952), the conditions (8.8)–(8.10) are appropriate only when the members of the frame obey the assumptions stated in Section 8.3.1, which lead to the form of (M, κ) relation shown in Fig. 8.6. If these assumptions are not obeyed, the conditions for shake-down to occur must be stated in terms of stress distributions across each section, rather than their moment resultants. However, the conditions (8.8)–(8.10), covering as they do the frequently occurring case of members with two axes of symmetry and equal yield stresses in tension and compression, are sufficient for many practical cases.

8.3.3 *Upper bound theorem*

The upper bound theorem is concerned with values of λ derived from assumed alternating plasticity or incremental collapse mechanisms. Alternating plasticity will first develop at the cross section in the frame at which the range of elastic bending moment $(\mathcal{M}^{max} - \mathcal{M}^{min})$ is greatest. If this cross section is denoted by k, the load factor at which alternating plasticity would be incipient is given by

$$\lambda'_a(\mathcal{M}_k^{max} - \mathcal{M}_k^{min}) = 2(M_y)_k. \tag{8.12}$$

The value λ'_a calculated in this way can be said to be the value of λ corresponding to an assumed alternating plasticity mechanism consisting of a single plastic hinge at section k.

If a particular incremental collapse mechanism is assumed, a corresponding value λ'_I of the load factor λ can be calculated. An example of this process was given in Section 8.2.3; it can be put in general terms as follows. Let θ_j denote the hinge rotation at cross section j during a small motion of the assumed mechanism. Positive and negative values of θ_j are distinguished by superscripts, θ_j^+ and θ_j^-. For the present purpose the assumed mechanism is treated as though it were the actual incremental collapse mechanism, with incremental collapse load factor λ'_I. Let m_j denote the residual bending moments in the frame at section j when it has shaken down. The value of m_j can be determined from one of the following two equations:

$$m_j + \lambda'_I.\mathcal{M}_j^{max} = (M_p)_j \quad \text{for all} \quad \theta_j^+. \tag{8.13}$$

$$m_j + \lambda'_I.\mathcal{M}_j^{min} = -(M_p)_j \text{ for all} \quad \theta_j^-. \tag{8.14}$$

Since the m_j are statically admissible with zero external loads, it follows from the Principle of Virtual Work that

$$\sum m_j\theta_j = 0, \tag{8.15}$$

where the summation covers all the hinge positions j in the assumed mechanism. Using Equations (8.13) and (8.14),

$$\sum [(M_p)_j - \lambda'_I.\mathcal{M}_j^{max}] \; \theta_j^+ + \sum [-(M_p)_j - \lambda'_I.\mathcal{M}_j^{min}] \; \theta_j^- = 0$$

$$\lambda'_I \sum [.\mathcal{M}_j^{max}\theta_j^+ + \mathcal{M}_j^{min} \; \theta_j^-] = \sum (M_p)_j \; |\theta_j|. \tag{8.16}$$

Equation (8.16) enables the value of λ'_I which corresponds to any assumed mechanism of incremental collapse to be calculated.

The upper bound theorem can now be stated as follows:

Upper bound theorem. The value of λ corresponding to any assumed mechanism of alternating plasticity (λ'_a) or of incremental collapse (λ'_I) must be either greater than or equal to the shake-down load factor λ_s.

Any value of λ calculated from an assumed mechanism will be referred to as satisfying the *kinematical conditions* of the upper bound theorem.

An upper bound theorem was developed by Koiter (1956, 1960). This theorem focuses attention on hypothetical cycles of plastic deformation. However, it does not provide a basis for a reasonably simple method of calculation for an upper bound on λ_s, whereas the upper bound theorem just stated is well suited to this purpose. The relationship between these two theorems has been clarified by Smith (1974).

8.3.4 *Observations on theorems*

The shake-down theorem is also valid for cases in which thermal stresses occur, as pointed out by Prager (1956). All that is necessary is to extend the definition of $\mathscr{M}$ to cover moments developed due to changes of temperature, assuming wholly elastic behaviour.

The presence of initial residual moments in a frame due to lack of fit of members, the fabrication process or the movement of supports has no influence on the conditions for shake-down to occur, and therefore on the value of λ_s. However, the elastic bending moment distribution depends upon joint and support stiffnesses, and so λ_s will also depend on these factors. This is in contrast to the situation in a plastic collapse analysis, since the value of the plastic collapse load factor λ_c is independent of joint and support stiffnesses, as pointed out in Section 2.6.

The shake-down theorem specifies conditions which, if satisfied, ensure that plastic flow will eventually cease. It does not enable upper bounds to be placed on the deflections which may develop in a frame subjected to variable repeated loading when λ does not exceed λ_s. Several attempts have been made to establish such bounds, for example by Capurso (1974), but none of these have been completely successful.

A uniqueness theorem for the value of λ_s can be formulated by combining the shake-down or lower bound theorem and the upper bound theorem. This theorem is stated for the sake of completeness:

Uniqueness theorem. If the statical conditions of the shake-down theorem and the kinematical conditions of the upper bound theorem are met, λ must be equal to the shake-down load factor λ_s.

8.4 Methods of analysis

A method of analysis suitable for simple beams and frames will now be described. It is based on the upper bound theorem, and consists essentially of determining the load factor λ_I' corresponding to each possible mechanism of incremental collapse. Each of these values is an upper bound on λ_s, and the load factor λ_a' for alternating plasticity is also an upper bound. The lowest of the upper bounds thus obtained will be the correct value of λ_s. The conclusion is finally checked by examining the statical conditions of the shake-down theorem.

8.4.1 *Illustrative calculation*

The frame used to illustrate the method is shown in Fig. 8.7(a). Each member has a plastic moment $25\,\text{kN}\,\text{m}$, with shape factor 1.15. The loads V and H can each vary independently between the following limits

$$V: \quad (16\lambda, 5\lambda)\text{kN}$$

$$H: \quad (10\lambda, 0)\text{kN}.$$

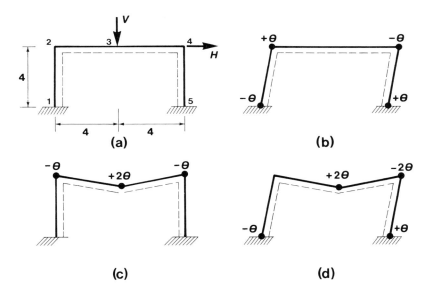

Fig. 8.7 *Rectangular portal frame and possible incremental collapse mechanisms*
  (a) Frame and loading
  (b) Sway mechanism
  (c) Beam mechanism
  (d) Combined mechanism

The first step is to determine the maximum and minimum elastic bending moments in the frame. The elastic bending moment distribution due to loads H and V is given in the first row of Table 8.5. At cross section 2, for example,

$$\mathcal{M}_2 = -0.8\,V + 0.75\,H.$$

Since the coefficients of V and H are negative and positive, respectively, the maximum elastic moment occurs at this section when V has its smallest value,

5λ, and H has its largest value, 10λ. Thus

$$\lambda\mathcal{M}_2^{\max} = -0.8\,(5\lambda) + 0.75\,(10\lambda) = 3.5\lambda,$$

the units being omitted for convenience. This value, together with the corresponding load combination $(5\lambda, 10\lambda)$ is entered in the table.

Table 8.5 *Elastic bending moments:* $V(16\lambda, 5\lambda); H(10\lambda, 0)$

Cross section	1	2	3	4	5
$\mathcal{M}$	$0.4V$ $-1.25H$	$-0.8V$ $+0.75H$	$1.2V$	$-0.8V$ $-0.75H$	$0.4V$ $+1.25H$
$\lambda\mathcal{M}^{\max}$	6.4λ $(16\lambda, 0)$	3.5λ $(5\lambda, 10\lambda)$	19.2λ $(16\lambda, -)$	-4λ $(5\lambda, 0)$	18.9λ $(16\lambda, 10\lambda)$
$\lambda\mathcal{M}^{\min}$	-10.5λ $(5\lambda, 10\lambda)$	-12.8λ $(16\lambda, 0)$	6λ $(5\lambda, -)$	-20.3λ $(16\lambda, 10\lambda)$	2λ $(5\lambda, 0)$
$\lambda(\mathcal{M}^{\max} - \mathcal{M}^{\min})$	16.9λ	16.3λ	13.2λ	16.3λ	16.9λ

The second and third rows of Table 8.5 give values of the maximum and minimum elastic moments at each section, calculated in this way, together with the load combinations which cause them. At section 3 the elastic moment is unaffected by H, and so no value of this load is specified in the two combinations.

The final row of the table gives values of the elastic bending moment range $\lambda(\mathcal{M}^{\max} - \mathcal{M}^{\min})$. The largest range is 16.9λ, occurring at sections 1 and 5; the alternating plasticity load factor λ_a' is therefore given by

$$16.9\lambda_a' = 2M_y = 50/1.15$$

$$\lambda_a' = 2.573.$$

In the analysis of the possible mechanisms of incremental collapse, the equations of equilibrium for residual moments will be required. These equations may be derived by using the principle of virtual displacements. For the three mechanisms shown in Fig. 8.7(b), (c) and (d), application of this principle leads to the equations

$$-m_1 + m_2 - m_4 + m_5 = 0. \tag{8.17}$$

$$-m_2 + 2m_3 - m_4 = 0. \tag{8.18}$$

$$-m_1 + 2m_3 - 2m_4 + m_5 = 0. \tag{8.19}$$

Equation (8.19) is not, of course, independent of the other two equations, from which it is obtained by addition.

The first possible mechanism of incremental collapse to be considered is the sway mechanism of Fig. 8.7(b). The plastic hinges shown would not, of course, all occur simultaneously. The residual moments at shake-down, having regard to the signs of the hinges, would be given by

$$m_1 + \lambda \mathcal{M}_1^{\min} = -M_p \qquad \theta_1 = -\theta.$$

$$m_2 + \lambda \mathcal{M}_2^{\max} = +M_p \qquad \theta_2 = +\theta.$$

$$m_4 + \lambda \mathcal{M}_4^{\min} = -M_p \qquad \theta_4 = -\theta.$$

$$m_5 + \lambda \mathcal{M}_5^{\max} = +M_p \qquad \theta_5 = +\theta.$$

With $M_p = 25$, and taking values of the elastic moments from Table 8.5,

$$
\begin{aligned}
m_1 - 10.5\lambda &= -25 \\
m_2 + 3.5\lambda &= +25 \\
m_4 - 20.3\lambda &= -25 \\
m_5 + 18.9\lambda &= +25.
\end{aligned}
\qquad (8.20)
$$

Substituting in Equation (8.17),

$$-(10.5\lambda - 25) + (-3.5\lambda + 25) - (20.3\lambda - 25) + (-18.9\lambda + 25) = 0$$

$$53.2\lambda = 100$$

$$\lambda = 1.880 = \lambda_I'.$$

This calculation follows the steps embodied in Equations (8.13)–(8.16). Assuming now the beam mechanism of Fig. 8.7(c),

$$
\begin{aligned}
m_2 - 12.8\lambda &= -25 & \theta_2 &= -\theta \\
m_3 + 19.2\lambda &= +25 & \theta_3 &= +2\theta \\
m_4 - 20.3\lambda &= -25 & \theta_4 &= -\theta.
\end{aligned}
\qquad (8.21)
$$

Substituting in Equation (8.18),

$$-(12.8\lambda - 25) + 2(-19.2\lambda + 25) - (20.3\lambda - 25) = 0$$

$$71.5\lambda = 100$$

$$\lambda = 1.399 = \lambda_I''.$$

Finally, for the combined mechanism of Fig. 8.7(d),

$$
\begin{aligned}
m_1 - 10.5\lambda &= -25 & \theta_1 &= -\theta \\
m_3 + 19.2\lambda &= +25 & \theta_3 &= +2\theta \\
m_4 - 20.3\lambda &= -25 & \theta_4 &= -2\theta \\
m_5 + 18.9\lambda &= +25 & \theta_5 &= +\theta.
\end{aligned}
\qquad (8.22)
$$

Substituting in Equation (8.19),

$$-(10.5\lambda - 25) + 2(-19.2\lambda + 25) - 2(20.3\lambda - 25) + (-18.9\lambda + 25) = 0$$

$$108.4\lambda = 150$$

$$\lambda = 1.384 = \lambda_I'''.$$

The upper bounds which have been derived are

alternating plasticity	λ_a'	$= 2.573$
incremental collapse: sway mechanism	λ_I'	$= 1.880$
beam mechanism	λ_I''	$= 1.399$
combined mechanism	λ_I'''	$= 1.384.$

From the upper bound theorem it is concluded that $\lambda_s = \lambda_I''' = 1.384$, failure being by incremental collapse in the combined mechanism.

This conclusion is now checked by verifying that the requirements of the uniqueness theorem are met. This involves determining the complete residual bending moment distribution in the frame after it has shaken down with $\lambda = \lambda_s = 1.384$. The values of m_1, m_3, m_4 and m_5 can be derived immediately from Equations (8.22) when $\lambda = 1.384$, and then m_2 is found from Equation (8.17) or (8.18). These residual moments are given in the first row of Table 8.6.

Table 8.6 *Bending moments after shake-down:* $\lambda_s = 1.384$

Cross section	1	2	3	4	5
m	-10.47	-6.23	-1.57	3.09	-1.15
$\lambda_s \mathcal{M}^{max}$	8.86	4.84	26.57	-5.54	26.15
$\lambda_s \mathcal{M}^{min}$	-14.53	-17.71	8.30	-28.09	2.76
M^{max}	-1.61	-1.39	25 $(16\lambda_s, -)$	-2.45	25 $(16\lambda_s, 10\lambda_s)$
M^{min}	-25 $(5\lambda_s, 10\lambda_s)$	-23.94	6.73	-25 $(16\lambda_s, 10\lambda_s)$	1.61
$M^{max} - M^{min}$	23.39	22.55	18.27	22.55	23.39

The remainder of the table is largely self-explanatory. The second and third rows are the maximum and minimum elastic moments for $\lambda = \lambda_s = 1.384$. In the next two rows, the maximum and minimum bending moments which could occur after the frame had shaken down are given; they are calculated as follows:

$$M^{\max} = m + \lambda_s \mathcal{M}^{\max}$$
$$M^{\min} = m + \lambda_s \mathcal{M}^{\min}$$

The final row of the table records the range of bending moment at each section, which may be compared with the available range of elastic behaviour $50/1.15$, or $43.48\,\text{kN m}$.

The statical conditions of the shake-down theorem, Equations (8.8)–(8.10), are all met by the distributions of bending moment given in Table 8.6. The kinematical conditions of the upper bound theorem are also satisfied, since the plastic moment is attained at sections 1, 3, 4 and 5. It follows from the uniqueness theorem that $\lambda_s = 1.384$.

The load combinations which would produce plastic hinges in the incremental collapse mechanism are given in Table 8.6. It will be seen that if λ exceeded λ_s incremental collapse would occur if H was held constant at 10λ while V varied cyclically between its extreme values 16λ and 6λ.

It is of interest to compare the value of λ_s which has been obtained with the plastic collapse load factor λ_c corresponding to a single application of the worst possible load combination. This combination is evidently

$$V = 16\lambda, \quad H = 10\lambda.$$

Failure by plastic collapse would occur with $\lambda_c = 1.442$, the collapse mechanism being the combined mechanism of Fig. 8.7(d). The value of λ_s, 1.384, is only 4 per cent below λ_c in this particular case.

It is to be expected that λ_s will never exceed the collapse load factor λ_c for the worst possible load combination, since the conditions of the shake-down theorem include those of the static theorem of plastic collapse as a special case. The factors governing the difference between λ_s and λ_c have been studied by Ogle (1964) and Heyman (1972).

8.4.2 Partial incremental collapse mechanism

A difficulty arises when the mechanism of incremental collapse has fewer than $(r + 1)$ plastic hinges, where r is the number of redundancies in the frame. To illustrate the point, consider again the frame of Fig. 8.7(a), with the loads now varying independently between the limits:

$$V: \quad (16\lambda, 5\lambda)\text{kN}$$
$$H: \quad (6\lambda, 0)\text{kN}.$$

The maximum and minimum elastic bending moments are then as given in Table 8.7.

Table 8.7 *Elastic bending moments:* $V(16\lambda, 5\lambda)$; $H(6\lambda, 0)$

Cross section	1	2	3	4	5
$\lambda \mathcal{M}^{max}$	6.4λ	0.5λ	19.2λ	−4λ	13.9λ
	(16λ, 0)	(5λ, 6λ)	(16λ, −)	(5λ, 0)	(16λ, 6λ)
$\lambda \mathcal{M}^{min}$	−5.5λ	−12.8λ	6λ	−17.3λ	2λ
	(5λ, 6λ)	(16λ, 0)	(5λ, −)	(16λ, 6λ)	(5λ, 0)
$\lambda(\mathcal{M}^{max} - \mathcal{M}^{min})$	11.9λ	13.3λ	13.2λ	13.3λ	11.9λ

Proceeding as in the previous example, the following upper bounds on λ_s are readily established:

$$\text{alternating plasticity} \qquad \lambda'_a = 3.269$$

$$\text{incremental collapse: sway mechanism} \qquad \lambda'_I = 2.688$$

$$\text{beam mechanism} \qquad \lambda''_I = 1.460$$

$$\text{combined mechanism} \quad \lambda'''_I = 1.623.$$

Evidently $\lambda_s = \lambda''_I = 1.460$, failure being by incremental collapse in the beam mechanism.

To verify this conclusion, it is necessary to show that the requirements of the uniqueness theorem can be met with $\lambda = \lambda_s = 1.460$. After shake-down has occurred at this value of λ, the residual moments at cross sections 2, 3 and 4, which are those involved in the incremental collapse mechanism, will be given by

$$m_2 + \lambda_s.\mathcal{M}_2^{min} = -M_p \qquad \theta_2 = -\theta$$

$$m_3 + \lambda_s.\mathcal{M}_3^{max} = +M_p \qquad \theta_3 = +2\theta$$

$$m_4 + \lambda_s.\mathcal{M}_4^{min} = -M_p \qquad \theta_4 = -\theta.$$

With $\lambda_s = 1.460$, and taking values of the elastic moments from Table 8.7,

$$m_2 = 12.8\lambda_s - 25 = -6.31$$

$$m_3 = 25 - 19.2\lambda_s = -3.03$$

$$m_4 = 17.3\lambda_s - 25 = 0.25.$$

These residual moments satisfy Equation (8.18). When substituted into Equation (8.17), it is found that

$$-m_1 + m_5 = 6.56 \qquad (8.23)$$

and these two residual moments cannot be determined uniquely. This is, of course, because the incremental collapse mechanism has one plastic hinge fewer than the $(r + 1)$ hinges required for a 'complete' mechanism.

Table 8.8 summarizes the position which has now been reached. The three residual moments m_2, m_3 and m_4 are known, and all the elastic moments have been evaluated with $\lambda = 1.460$.

Table 8.8 *Bending moments after shake-down:* $\lambda_s = 1.460$

Cross section	1	2	3	4	5
m		-6.31	-3.03	0.25	
$\lambda_s . \mathscr{M}^{max}$	9.34	0.73	28.03	-5.83	20.29
$\lambda_s . \mathscr{M}^{min}$	-8.03	-18.69	8.76	-25.25	2.92
M^{max}		-5.58	25 $(16\lambda_s, -)$	-5.58	
M^{min}		-25 $(16\lambda_s, 0)$	5.73	-25 $(16\lambda_s, 6\lambda_s)$	
$M^{max} - M^{min}$	17.37	19.42	19.27	19.42	17.37

To satisfy the requirements of the shake-down theorem, m_1 and m_5 must satisfy Equation (8.23) and also comply with the following conditions

$$m_1 + \ 9.34 \leqslant \ 25$$
$$m_1 - \ 8.03 \geqslant -25$$
$$m_5 + 20.29 \leqslant \ 25$$
$$m_5 + \ 2.92 \geqslant -25.$$

Taking these inequalities in pairs, the following two continuing inequalities are obtained:

$$-16.97 \leqslant m_1 \leqslant 15.66. \tag{8.24}$$
$$-22.08 \leqslant m_5 \leqslant \ 4.71. \tag{8.25}$$

From Equation (8.23),

$$m_5 = m_1 + 6.56,$$

and combining this with the inequalities (8.25)

$$-28.64 \leqslant m_1 \leqslant -1.85. \tag{8.26}$$

Taking the more stringent requirements of (8.24) and (8.26),

$$-16.97 \leqslant m_1 \leqslant -1.85. \tag{8.27}$$

Thus any value of m_1 lying between -16.97 and -1.85, together with the corresponding value of m_5 from the equilibrium equation (8.23), will satisfy the requirements of the shake-down theorem. One such pair of values is

$$m_1 = -10, \quad m_5 = -3.44.$$

In this particular case a suitable pair of values of m_1 and m_5 could easily have been determined by inspection.

If Table 8.8 is completed using, for example, the values of m_1 and m_5 given above, it will be seen that all the requirements of the uniqueness theorem are met, thus confirming that $\lambda_s = 1.460$. The cycle of loading which would cause incremental collapse if λ exceeded λ_s is seen from the entries in Table 8.8 to consist of V being held constant at 16λ while H varied between its extreme values of 6λ and zero.

Under the worst possible combination of loads, namely

$$V = 16\lambda, \quad H = 6\lambda,$$

the plastic collapse load factor λ_c is 1.563, corresponding to the beam mechanism. The value of λ_s, 1.460, is 6.6 per cent below this value of λ_c.

8.4.3 *Other methods of calculation*

The method which has just been outlined comprises two steps. The first of these is the determination of the value of λ_s corresponding to every possible mechanism of incremental collapse, together with the value of λ_a'. Once the correct mechanism of incremental collapse has been established (assuming that failure does not occur by alternating plasticity), the second step follows; this involves confirming that all the requirements of the uniqueness theorem can be met when $\lambda = \lambda_s$, as in Tables 8.6 and 8.8.

If the first of these steps is circumvented in some way, a trial-and-error procedure of the kind suggested by Neal and Symonds (1950) results. For instance, a plastic collapse analysis might first be carried out for the worst combination of loads, leading to the plastic collapse mechanism; it would then be a reasonable assumption that incremental collapse would occur by the same mechanism. An analysis of this mechanism of incremental collapse would then be performed, leading to a table such as 8.6 or 8.8. The assumed mechanism would be correct unless one or more of the calculated values of $M^{\max}$ or $M^{\min}$ exceeded the plastic moment in magnitude.

The concepts underlying the method of combining mechanisms for plastic collapse were extended by Symonds and Neal (1951) to cover the case of incremental collapse. However, it is more difficult to judge which combinations of mechanisms are likely to lead to the lowest corresponding value of λ_s, and so the technique is less effective for incremental collapse than for plastic collapse.

A lower bound approach was first developed by Symonds and Neal (1950). This was based on solving the inequalities (8.8)–(8.10) contained in the shake-down theorem, taking into account the conditions of equilibrium which must be obeyed by the residual moments. Heyman (1959) showed that this approach is particularly suitable for solution using a digital computer program. The design of frames for minimum weight if the criterion of design is the provision of a prescribed load factor against incremental collapse or alternating plasticity has been discussed by Heyman (1951, 1958).

8.4.4 *Estimates of deflections*

It is interesting to note that if certain quite plausible assumptions are made, an estimate can be made of the deflections in a frame after it has shaken down. Suppose that if λ exceeded λ_s, failure would occur by incremental collapse rather than alternating plasticity, so that $\lambda_s = \lambda_I$. It is then assumed that when the frame is subjected to a large number of the critical loading cycles with $\lambda = \lambda_I$, there will be one cross section, say q, at which M^{max} or M^{min} tends asymptotically towards the plastic moment in magnitude, but no plastic hinge rotation actually takes place at this section. Meanwhile, plastic hinges will have formed and undergone rotation at all the other sections involved in the incremental collapse mechanism. Only if λ exceeded λ_I would there be any plastic hinge rotation at section q.

This assumption was obeyed by the rectangular frame of Fig. 8.1 which was analysed in Section 8.2.3. In this case the incremental collapse mechanism involved plastic hinges at sections 1, 3, 4 and 5. Table 8.4 shows that the effect of repetition of the critical loading cycle with $W = W_I$ (equivalent to $\lambda = \lambda_I$) was to cause M_3^{max} to approach M_p asymptotically, but there was no plastic hinge rotation at this section.

For the shaken down condition of a frame with $\lambda = \lambda_I$, the residual bending moment distribution may be determined, as in Sections 8.4.1 and 8.4.2. With the assumption of continuity at one of the hinge positions in the incremental collapse mechanism, the technique described in Section 5.4 can be used to identify this position. The plastic hinge rotations at the other sections, and also the deflections of the frame when free from load, can then be determined.

8.5 Relation to design

As pointed out in Section 2.7, the plastic design of steel frames is appropriate when plastic collapse is the ultimate limit state which governs the design, and this is the situation in many practical cases. However, if a frame is subjected to variable repeated loading, shake-down will not occur if the load factor exceeds a value λ_s, which is less than the load factor λ_c at which plastic collapse would occur under the worst possible combination of loads. This raises the question as to whether the indefinite continuation of plastic flow, by either alternating plasticity or incremental collapse, is a more relevant ultimate limit state than plastic collapse for this type of loading. However, Horne (1954) has shown that this is unlikely to be the case. A frame designed by the plastic method will have an acceptably small probability of failure by plastic collapse; the probability of the occurrence of either alternating plasticity or incremental collapse will almost certainly be much less. Plastic collapse is the appropriate limit state in these circumstances.

There are several points to be considered when comparing these ultimate limit states. The principal one is that plastic collapse only requires a single application of the appropriate load combination with $\lambda = \lambda_c$. The probability of this occurrence will be denoted by p. By contrast, incremental collapse requires the application of several cycles of loading at a load factor greater than λ_I. Each cycle of loading will include two or more combinations of load, so that unacceptably large deflections would only develop after, say, n applications of load, where n is of the order of 10. If the probability attached to a single application of one of these combinations of load at a load factor greater than λ_I is q, the probability of a failure by incremental collapse will be q^n. Since λ_I is less than λ_c, q will be greater than p. However, q will generally be small, so that q^n is likely to be much less than p. This argument is put rather crudely; variable repeated loading will usually be stochastic in nature. Nevertheless, the essential point, namely that incremental collapse requires a number of load applications, to each of which is attached a small probability, will be clear.

The ratio λ_I/λ_c obviously has a bearing on the argument just outlined, which depends essentially upon q being small. For the two examples given in Section 8.4, this ratio was 0.96 and 0.934. These are fairly typical values, and ratios less than 0.8 are rarely encountered. It is therefore reasonable to assume that q is small, unless the frame has unusual proportions or is subjected to an abnormal type of loading.

A further point is that incremental collapse only occurs when λ exceeds λ_I, and the increments of deflection which take place in each cycle become larger the more λ exceeds λ_I. For incremental collapse to occur as a result of a reasonably small number of applications of load, say $n = 10$, λ will need to exceed λ_I by an appreciable margin, and this makes it even more likely that q is small.

Finally, it will be appreciated that a failure by incremental collapse is gradual, the deflections building up over a period, so that ample warning of the progress of this type of failure will usually be available. A greater likelihood of this kind of failure could therefore be tolerated than for plastic collapse under a single combination of loads, for which there would be no prior indications of the imminence of failure.

Alternating plasticity, if it occurs, does not cause the growth of large deflections. The only risk involved is of fracture due to low endurance fatigue. Various investigators, for example Royles (1966), have shown that the life of mild steel beams subjected to large reversals of strain, several times the yield strain, is of the order of $10^2 - 10^4$ cycles. The implication is that alternating plasticity is most unlikely to be a relevant ultimate limit state. If a large number of cycles of load of the order of $10^5 - 10^7$ could occur, fatigue would become a dominant consideration in the design.

References

Bleich, H. (1932), 'Über die Bemessung statisch unbestimmter Stahltragwerke unter Berücksichtigung des elastisch-plastischen Verhaltens des Baustoffes', *Bauingenieur*, **13**, 261.

Capurso, M. (1974), 'A displacement bounding principle in shakedown of structures subjected to cyclic loads', *Int. J. Solids Structures*, **10**, 77.

Gozum, A. and Haaijer, G. (1955), 'Deflection stability (shakedown) of beams', Fritz Eng. Lab. Report no. 205 G.1.

Grüning, M. (1926), *Die Tragfähigkeit statisch unbestimmten Tragwerke aus Stahl bei beliebig haufig wiederholter Belastung*, Julius Springer, Berlin.

Heyman, J. (1951), 'Plastic design of beams and plane frames for minimum material consumption', *Q. Appl. Math.*, **8**, 373.

Heyman, J. (1958), 'Minimum weight design of frames under shakedown loading', *J. Eng. Mech. Div., Proc. Am. Soc. Civil Engrs*, **84**, (EM 4), Paper 1790.

Heyman, J. (1959), 'Automatic analysis of steel framed structures under fixed and varying loads', *Proc. Inst. Civil Engrs*, **12**, 39.

Heyman, J. (1972), 'The significance of shakedown loading', prelim. publ., 9th Congr. Int. Assoc. Bridge Struct. Eng., Amsterdam, 1972.

Horne, M. R. (1949), 'The effect of variable repeated loads in the plastic theory of structures', *Research (Eng. Struct. Suppl.), Colston Papers*, **2**, 141.

Horne, M. R. (1954), 'The effect of variable repeated loads in building structures designed by the plastic theory', *Proc. Int. Assoc. Bridge Struct. Eng.*, **14**, 53.

Kazinczy, G. V. (1931), 'Die Weiterentwicklung der Elastizitätstheorie', *Technika*, Budapest.

Koiter, W. T. (1952), 'Some remarks on plastic shakedown theorems', 8th Int. Congr. Theor. Appl. Mech., Istanbul, 1952.

Koiter, W. T. (1956), 'A new general theorem on shake-down of elastic-plastic structures', *Proc. k. Ned. Akad. Wet.* (B), **59**, 24.

Koiter, W. T. (1960), 'General theorems for elastic-plastic solids', *Progress in Solid Mechanics*, vol. I, Amsterdam.

Massonet, C. (1953), 'Essais d'adaptation et de stabilisation plastiques sur les poutrelles laminées', *Proc. Int. Assoc. Bridge Struct. Eng.*, **13**, 239. (See also *Ossat. métall.*, **19**, 318, 1954.)

Melan, E. (1936), 'Theorie statisch unbestimmter Systeme', prelim. publ., 2nd Congr. Int. Assoc. Bridge Struct. Eng., 43, Berlin.

Neal, B. G. (1950), 'Plastic collapse and shakedown theorems for structures of strain-hardening material', *J. Aero. Sci.*, **17**, 297.

Neal, B. G. (1951), 'The behaviour of framed structures under repeated loading', *Q. J. Mech. Appl. Math.*, **4**, 78.

Neal, B. G. and Symonds, P. S. (1950), 'A method for calculating the failure load for a framed structure subjected to fluctuating loads', *J. Inst. Civil Engrs,* **35**, 186.

Neal, B. G. and Symonds, P. S. (1958), 'Cyclic loading of portal frames: theory and tests', *Proc. Int. Assoc. Bridge Struct. Eng.,* **18**, 171. (See also Symonds, P. S. (1953), 'Cyclic loading tests on small frames', Final Report, 4th Congr. Int. Assoc. Bridge Struct. Eng., Cambridge, 1953, 109.

Ogle, M. H. (1964), 'Shakedown of steel frames', Ph.D. thesis, Cambridge Univ.

Popov, E. P. and McCarthy, R. E. (1960), 'Deflection stability of frames under repeated loads', *J. Eng. Mech. Div., Proc. Am. Soc. Civil Engrs,* **86**, (EM 1), 61.

Prager, W. (1956), 'Plastic design and thermal stresses', *B. Weld. J.,* **3**, 355.

Royles, R. (1966), 'Low endurance fatigue behaviour of mild steel beams in reversed bending', *J. Strain Anal.,* **1**, 239.

Smith, D. L. (1974), 'Plastic limit analysis and synthesis of structures by linear programming', Ph.D. thesis, London Univ.

Symonds, P. S. and Neal, B. G. (1950), 'The calculation of failure loads on plane frames under arbitrary loading programmes', *J. Inst. Civil Engrs,* **35**, 41.

Symonds, P. S. and Neal, B. G. (1951), 'Recent progress in the plastic methods of structural analysis', *J. Franklin Inst.,* **252**, 383, 469.

Symonds, P. S. and Prager, W. (1950), 'Elastic-plastic analysis of structures subjected to loads varying arbitrarily between prescribed limits', *J. Appl. Mech.,* **17**, 315.

Examples

1. A uniform beam of T-section has width a and depth $1.24a$. Its cross section consists of two similar rectangles whose sides are a and $0.24a$, and the material is ideal plastic (see Fig. 1.3(b)). A bending moment is applied about an axis parallel to the flange, causing flexure in the plane of the web. Find the position of the neutral axis if the beam behaves elastically. If the bending moment is increased until the whole section is fully plastic, show that the neutral axis moves parallel to itself to the equal area axis through a distance $0.19a$.

If the bending moment is then reduced from the plastic moment, show that the neutral axis immediately assumes a new position in the web at a distance $0.2a$ from the equal area axis so that the stress remains constant at the yield stress over a portion of the web of this length. Hence show that upon unloading from the plastic moment the elastic range of bending moment is 0.36 per cent greater than for the initially unstressed beam, whereas the flexural rigidity is 0.88 per cent less.

2. A uniform beam ABCDE whose plastic moment is 30 kN m is simply supported at A, C and E and carries concentrated loads P and Q at B and D, respectively:

$$AB = BC = CD = DE = 2m.$$

The loads can vary independently between the limits

$$P(30\lambda, 0) \quad Q(20\lambda, 0) \text{kN}$$

Find the values of λ_s and λ_c, assuming a shape factor 1.15. The elastic bending moments due to P and Q are

$$\text{B: } (13P - 3Q)/16 \text{ kN m}$$

$$\text{C: } -(6P + 6Q)/16$$

$$\text{D: } (-3P + 13Q)/16,$$

sagging bending moments being reckoned positive.

3. A uniform beam ABCDEFG whose plastic moment is 40 kN m is simply supported at A, C, E and G and carries concentrated loads P, Q and R at B, D and F, respectively:

$$\text{AB} = \text{BC} = \text{CD} = \text{DE} = \text{EF} = \text{FG} = 2\text{m}.$$

The loads can vary independently between the limits

$$P(40\lambda, 0) \quad Q(20\lambda, 0) \quad R(20\lambda, 0) \text{kN}.$$

Find the values of λ_s and λ_c, assuming a shape factor of 1.15. The elastic bending moments due to P, Q and R are

$$\text{B: } 0.8P - 0.15Q + 0.05R \text{ kN m}$$

$$\text{C: } -0.4P - 0.3Q + 0.1R$$

$$\text{D: } -0.15P + 0.7Q - 0.15R$$

$$\text{E: } 0.1P - 0.3Q - 0.4R$$

$$\text{F: } 0.05P - 0.15Q + 0.8R,$$

sagging bending moments being reckoned positive.

4. A uniform beam ABCD whose plastic moment is 36 kN m is rigidly built-in at A and simply supported at D. It carries concentrated loads P and Q at B and C, respectively.

$$\text{AB} = \text{BC} = \text{CD} = 1.5\text{m}.$$

The loads can vary independently between the limits

$$P(20\lambda, 0) \quad Q(20\lambda, 0) \text{kN}.$$

Find the values of λ_s and λ_c, assuming a shape factor of unity. The elastic bending moments due to P and Q are:

$$\text{A: } (-15P - 12Q)/18 \text{ kN m}$$

$$\text{B: } (8P + Q)/18$$

C: $(4P + 14Q)/18$,

sagging bending moments being reckoned positive.

5. A uniform beam ABCD whose plastic moment is 45 kN m is rigidly built-in at A and D, and carries concentrated loads P and Q at B and C, respectively:

$$AB = BC = CD = 3m.$$

The loads can vary independently between the limits

$$P(20\lambda, 0) \quad Q(20\lambda, -20\lambda)kN$$

Find the values of λ_s and λ_c, assuming a shape factor of unity. The elastic bending moments due to P and Q are:

A: $(-12P - 6Q)/9$ kN m

B: $(8P + Q)/9$

C: $(P + 8Q)/9$

D: $(-6P - 12Q)/9$,

sagging bending moments being reckoned positive.

6. A uniform beam AB of length l is built-in at both ends and has a plastic moment M_p. It carries a concentrated load W which rolls back and forth along the beam.

Plot the elastic bending moment diagram for various positions of the load, and hence construct a diagram showing the maximum and minimum elastic bending moments at each section. Hence determine W_s, and compare its value with W_c.

If the concentrated load W is at a position C, distance μl from A, the elastic bending moments are:

A: $-Wl\mu (1 - \mu)^2$

C: $2Wl\mu^2 (1 - \mu)^2$

B: $-Wl\mu^2 (1 - \mu)$,

sagging bending moments being reckoned positive.

7. A fixed-base rectangular portal frame ABCDE consists of two columns, AB and ED, each of length 3.5 m, and a beam BD also of length 3.5 m. The frame is of uniform section throughout, with plastic moment 40 kN m and shape factor 1.12. The beam carries a vertical concentrated load V at its mid-point C, and a horizontal concentrated load H is applied at D in the direction BD. The loads can vary independently between the limits:

$$V(48\lambda, 0) \quad H(24\lambda, -12\lambda)kN$$

Find the value of λ_s.

The elastic bending moments, using the sign convention of Fig. 8.1, are

$$A: -H + 7V/48 \text{ kN m}$$

$$B: 0.75H - 7V/24$$

$$C: 7V/12$$

$$D: -0.75H - 7V/24$$

$$E: H + 7V/48.$$

8. Find the value of λ_s for the frame of example 7 if the loads can vary independently between the limits

$$V(48\lambda, 0) \quad H(20\lambda, -12\lambda) \text{kN.}$$

9. In the uniform, fixed-base frame shown in Fig. 8.1, all the members have a plastic moment M_p and a shape factor of unity. The loads can vary independently between the limits

$$V(2W, 0) \quad H(W, 0).$$

Find the value W_s of W above which incremental collapse would occur. The elastic bending moments due to V and H are given in Equations (8.1).

10. For the frame of example 9, estimate the deflection h corresponding to the load H when shake-down has occurred after a large number of cycles of loading with $W = W_s$ and the loads have then been removed from the frame. The effects of strain-hardening and the spread of plastic zones along the members may be neglected.

The compatibility equations (2.19)–(2.21), derived in Section 2.5.3, and the expression for h given in Equation (2.24) may be used.

Hint: Show that there must be some plastic hinge rotation at section 5 before shake-down.

APPENDIX A
Proofs of Plastic Collapse Theorems

Constancy of curvatures during plastic collapse. A state of plastic collapse is defined as one in which the deflections of the frame continue to increase while the loads remain constant. From this definition it can be shown that during collapse the distributions of bending moment and of curvature in the frame remain unaltered as the deflections increase. To prove this, consider the changes which occur during a definite small interval of time in which plastic collapse is occurring. These are denoted by the prefix δ, and are

$\delta M, \delta \kappa$: changes of bending moment and curvature at any cross section other than a plastic hinge position.

$\delta M_i, \delta \theta_i$: changes of bending moment and hinge rotation at plastic hinge position i.

Since the loads remain unchanged, the changes of bending moment δM and δM_i must satisfy the conditions of equilibrium with zero external load. The changes of curvature and hinge rotation $\delta \kappa$ and $\delta \theta_i$ must be compatible. It follows from the Principle of Virtual Work, Equation (2.11), that

$$\int \delta M \, \delta \kappa \, ds + \sum \delta M_i \delta \theta_i = 0, \tag{A1}$$

where the integral covers all the members of the frame and the summation covers all the plastic hinge positions.

The plastic hinge hypothesis is that hinge rotation only takes place when the bending moment remains constant at the plastic moment, so that $\delta M_i = 0$. It follows that each term $\delta M_i \, \delta \theta_i$ in Equation (A1) must be zero, so that

$$\int \delta M \delta \kappa \, ds = 0. \tag{A2}$$

Provided that the (M, κ) relation is such that increments of bending moment and curvature are always of the same sign,

$$\delta M \delta \kappa \geqslant 0. \tag{A3}$$

It then follows from Equation (A2) that δM and $\delta \kappa$ must both be zero at every cross section. Thus during plastic collapse the bending moment and curvature distributions remain unaltered. The increases of deflection which occur are due solely to the rotations which occur at the plastic hinges, which must therefore constitute a mechanism motion.

The proviso embodied in (A3) is important. It precludes from consideration a strain-softening type of behaviour in which an increase in the magnitude of the curvature is accompanied by a decrease in the magnitude of the bending moment.

Static theorem. The static theorem was stated in Section 3.2.1 as follows: If there exists any distribution of bending moment throughout a frame which is both safe and statically admissible with a set of loads λ, the value of λ must be less than or equal to the collapse load factor λ_c.

Let the actual collapse mechanism involve changes of plastic hinge rotation $\delta\theta_i$ at each plastic hinge position i, and also changes of displacement corresponding to each characteristic load P_r which are denoted by δd_r.

Let M_i' denote the bending moment at cross section i in a distribution of bending moment which is both safe and statically admissible with the set of loads λ, in accordance with the theorem. Further, let M_i denote the bending moment at section i in the actual distribution of bending moment at collapse. This distribution must be safe and statically admissible with the set of loads λ_c.

Using the Principle of Virtual Work,

$$\sum \lambda P_r\, \delta d_r = \sum M_i'\, \delta\theta_i$$

$$\sum \lambda_c P_r \delta d_r = \sum M_i \delta\theta_i.$$

In these equations the summations on the left-hand side cover all points of application of load, and those on the right-hand side cover all plastic hinge positions.

These two equations can be combined to give

$$\lambda_c \sum M_i'\, \delta\theta_i = \lambda \sum M_i\, \delta\theta_i. \tag{A4}$$

Whenever $\delta\theta_i$ is positive, M_i will be equal to the plastic moment $+(M_p)_i$. Since M_i' is safe it cannot exceed $+(M_p)_i$. It follows that

$$M_i\, \delta\theta_i \geqslant M_i'\, \delta\theta_i \quad \text{if} \quad \delta\theta_i > 0.$$

Similarly, if $\delta\theta_i$ is negative, $M_i = -(M_p)_i$, and $M_i' \geqslant -(M_p)_i$, so that again

$$M_i\, \delta\theta_i \geqslant M_i'\, \delta\theta_i \quad \text{if} \quad \delta\theta_i < 0.$$

It follows at once from Equation (A4) that

$$\lambda \leqslant \lambda_c,$$

which establishes the theorem.

Kinematic theorem. The kinematic theorem was stated in Section 3.2.2 as follows: For a given frame subjected to a set of loads λ, the value of λ which corresponds to any assumed mechanism must be either greater than or equal to the collapse load factor λ_c.

Let the assumed mechanism involve changes of plastic hinge rotation $\delta\theta'_k$ at each plastic hinge position k, and also changes of displacement corresponding to each characteristic load P_r which are denoted by $\delta d'_r$.

Let M'_k denote the plastic moment at cross section k, with a sign corresponding to that of the plastic hinge rotation, so that

$$M'_k = +(M_p)_k \quad \text{if} \quad \delta\theta'_k > 0$$
$$M'_k = -(M_p)_k \quad \text{if} \quad \delta\theta'_k < 0. \tag{A5}$$

The value of λ corresponding to the assumed mechanism is then obtained from the equation

$$\sum \lambda P_r \, \delta d'_r = \sum M'_k \, \delta\theta'_k. \tag{A6}$$

If M_k denotes the bending moment at section k in the actual distribution of bending moment at collapse, it follows from the Principle of Virtual Work that

$$\sum \lambda_c P_r \delta d'_r = \sum M_k \, \delta\theta'_k. \tag{A7}$$

Combining equations (A6) and (A7),

$$\lambda_c \sum M'_k \, \delta\theta'_k = \lambda \sum M_k \delta\theta'_k. \tag{A8}$$

Whenever $\delta\theta'_k$ is positive, $M'_k = +(M_p)_k$, from (A5). Since M_k is safe it cannot exceed $+(M_p)_k$. It follows that

$$M'_k \, \delta\theta'_k \geqslant M_k \, \delta\theta'_k \quad \text{if} \quad \delta\theta'_k > 0.$$

Similarly, if $\delta\theta'_k$ is negative, $M'_k = -(M_p)_k$, and $M_k \geqslant -(M_p)_k$, so that again

$$M'_k \, \delta\theta'_k \geqslant M_k \, \delta\theta'_k \quad \text{if} \quad \delta\theta'_k < 0.$$

It follows at once from Equation (A8) that

$$\lambda \geqslant \lambda_c,$$

which establishes the theorem.

Proofs of Shake-down Theorems

Shake-down or lower bound theorem. The shake-down theorem was stated in Section 8.3.2 as follows: If there exists any distribution of residual bending moment $\bar{m}$ throughout a frame which is statically admissible with zero external loading and which also satisfies at every cross section j the conditions

$$\bar{m}_j + \lambda \,\mathcal{M}_j^{\max} \leqslant (M_p)_j \tag{8.8}$$

$$\bar{m}_j + \lambda \,\mathcal{M}_j^{\min} \geqslant -(M_p)_j \tag{8.9}$$

$$\lambda(\mathcal{M}_j^{\max} - \mathcal{M}_j^{\min}) \leqslant 2(M_y)_j \tag{8.10}$$

the value of λ will be less than or equal to the shake-down load factor λ_s.

This theorem will be proved for the case in which all the members of the frame have a shape factor of unity, so that $M_y = M_p$. In this case the inequalities (8.10) may be discarded, as they are contained within (8.8) and (8.9). This assumption implies that the (M, κ) relation for each member is the ideal relation of Fig. 2.1, for which

$$M = M_p, \qquad \delta\theta > 0$$

$$M = -M_p, \qquad \delta\theta < 0$$

$$|M| < M_p, \qquad \delta M = EI\delta\kappa.$$

The theorem is proved by considering the positive definite quantity U, defined by

$$U = \int \frac{(m_j - \bar{m}_j)^2}{2(EI)_j}\,ds_j. \tag{B1}$$

In this equation m_j represents the actual residual moment at cross section j during any stage of the loading, and $\bar{m}_j$ is a distribution of bending moment satisfying the conditions (8.8) and (8.9). $(EI)_j$ and ds_j are the flexural rigidity and element of length respectively at the section j, and the integration covers all the members of the frame.

Suppose now that during a definite small interval of time there are small changes in the applied loads, causing changes which will be denoted by the prefix δ. From Equation (B1),

$$\delta U = \int (m_j - \bar{m}_j) \frac{\delta m_j}{(EI)_j} \, ds_j. \tag{B2}$$

It will now be shown that δU is always negative.

As pointed out in Section 8.3.1, residual moments are defined for the purpose of shake-down analysis by the equation

$$m_j = M_j - \mathcal{M}_j \tag{8.7}$$

which in incremental form is

$$\delta m_j = \delta M_j - \delta \mathcal{M}_j. \tag{B3}$$

If there are changes in plastic hinge rotation $\delta\theta_k$ at cross sections denoted by k, these will be compatible with the actual changes of curvature $\delta M_j/(EI)_j$. The changes of curvature $\delta\mathcal{M}_j/(EI)_j$ which would have occurred if the entire frame had responded elastically to the same small changes of load must satisfy the requirements of compatibility with zero changes of plastic hinge rotation. It follows that the changes of curvature $(\delta M_j - \delta\mathcal{M}_j)/(EI)_j$, which from Equation (B3) are equal to $\delta m_j/(EI)_j$, must be compatible with the plastic hinge rotations $\delta\theta_k$. Using these compatible changes of curvature and hinge rotation in the virtual work equation, together with the distribution of residual moments $(m_j - \bar{m}_j)$, which must be statically admissible with zero external loads, it is found that

$$\int (m_j - \bar{m}_j) \frac{\delta m_j}{(EI)_j} \, ds_j + \sum (m_k - \bar{m}_k)\delta\theta_k = 0. \tag{B4}$$

From Equation (B2), it follows that

$$\delta U = -\sum (m_k - \bar{m}_k)\delta\theta_k. \tag{B5}$$

Suppose now that at a particular section k,

$$(m_k - \bar{m}_k) < 0. \tag{B6}$$

Using the inequality (8.8)

$$m_k < \bar{m}_k \leqslant (M_p)_k - \lambda \mathcal{M}_k^{\max}$$

$$m_k + \lambda \mathcal{M}_k^{\max} < (M_p)_k.$$

This result shows that the plastic hinge which is undergoing rotation at this section cannot be of positive sign, so that $\delta\theta_k < 0$. From Equation (B6) it follows that

$$(m_k - \bar{m}_k)\delta\theta_k > 0. \tag{B7}$$

By a similar argument it can be shown that if $(m_k - \bar{m}_k) > 0$, $\delta\theta_k$ must be positive, so that the inequality (B7) also holds true in this case. It can therefore be concluded that

$$(m_k - \bar{m}_k)\delta\theta_k \geqslant 0, \tag{B8}$$

the equality sign covering those sections where $m_k = \bar{m}_k$.

Combining this condition with Equation (B5), it is seen that

$$\delta U \leqslant 0. \tag{B9}$$

It is evident from Equation (B5) that δU will be zero if no plastic hinge rotation occurs during the interval considered, for the $\delta\theta_k$ will then all be zero. Thus U decreases whenever any plastic hinges undergo rotation and remains constant when the behaviour is wholly elastic. Since U is positive definite, it must either eventually become zero, in which case the distributions m_j and $\bar{m}_j$ would be identical, or else settle down at some positive value and thereafter remain unchanged. In either case the frame would have shaken down, thus establishing the theorem.

Upper bound theorem. The upper bound theorem was stated in Section 8.3.3 as follows: The value of λ corresponding to any assumed mechanism of alternating plasticity (λ'_a) or of incremental collapse (λ'_I) must be either greater than or equal to the shake-down load factor λ_s.

The first part of this theorem, that $\lambda'_a \geqslant \lambda_s$, is virtually self-evident. Shakedown could not occur if the elastic range of bending moment at any cross section, $\lambda(\mathcal{M}^{\max} - \mathcal{M}^{\min})$ exceeded the available range of elastic response $2M_p$. Thus

$$\lambda_s(\mathcal{M}^{\max} - \mathcal{M}^{\min}) \leqslant 2M_p = \lambda'_a(\mathcal{M}^{\max} - \mathcal{M}^{\min}),$$

so that

$$\lambda_s \leqslant \lambda'_a.$$

The load factor λ'_I corresponding to an assumed incremental collapse mechanism is calculated from Equation (8.16), as follows:

$$\lambda'_I \sum [\mathcal{M}_k^{\max}\theta_k^+ + \mathcal{M}_k^{\min}\theta_k^-] = \sum (M_p)_k |\theta_k|. \tag{8.16}$$

In this equation the hinge rotations are denoted by θ_k, and on the left-hand side superscripts are used to indicate the sign of each hinge.

From the shake-down theorem, condition (8.8) gives

$$\lambda_s \mathcal{M}_k^{\max} \leqslant (M_p)_k - \bar{m}_k.$$

At each hinge position where the rotation is positive, it follows that

$$\lambda_s \mathcal{M}_k^{\max} \theta_k^+ \leqslant (M_p)_k\theta_k^+ - \bar{m}_k\theta_k^+. \tag{B10}$$

Similarly, condition (8.9) of the shake-down theorem implies that

$$\lambda_s \mathcal{M}_k^{\min} \geqslant -(M_p)_k - \bar{m}_k$$

so that at each hinge position where the rotation is negative,

$$\lambda_s \mathcal{M}_k^{\min} \theta_k^- \leqslant -(M_p)_k \theta_k^- - \bar{m}_k \theta_k^-. \tag{B11}$$

Using the inequalities (B10) and (B11), and summing over all the plastic hinge positions in the assumed mechanism, it is found that

$$\lambda_s \sum [\mathcal{M}_k^{\max} \theta_k^+ + \mathcal{M}_k^{\min} \theta_k^-] \leqslant \sum (M_p)_k |\theta_k| - \sum \bar{m}_k \theta_k. \tag{B12}$$

Since the residual bending moments $\bar{m}_k$ are statically admissible with zero external load, it follows from the Principle of Virtual Work that

$$\sum \bar{m}_k \theta_k = 0 \tag{B13}$$

and therefore

$$\lambda_s \sum [\mathcal{M}_k^{\max} \theta_k^+ + \mathcal{M}_k^{\min} \theta_k^-] \leqslant \sum (M_p)_k |\theta_k|. \tag{B14}$$

Comparing this result with Equation (8.16), it follows at once that

$$\lambda_s \leqslant \lambda_I',$$

which establishes the theorem.

Answers to Examples

Chapter 1

2. $1848 \, cm^3$ 3. 1.80 4. $0.1B$ from centre, $0.3B^3 \sigma_0$
5. 0.741 6. $0.6D$, $0.32\sigma_0$
7. $0.667M_p/l$, $0.5M_p/l$, $0.364M_p/l$
8. $1.5B^2 T\sigma_0$, $\dfrac{M_x}{M_p} = 1 - \dfrac{3}{4}\left(\dfrac{M_y}{M_p}\right)^2$

Chapter 2

1. $8M_p/l$, $4.5M_p/l$ 2. $6M_p/l$, $7.5M_p/l$, $9M_p/l$
3. $5.33M_p/l$ 4. $4M_p/l$ 5. $2M_p/\mu(1-\mu)l$

Chapter 3

1. 1.6; 10.67, 18.67 kN   2. $(6 + 4\sqrt{2})M_p/l$
3. 12.87, 23.36, 15.44 kN m   4. $576M_p/49l$, $9M_p/l$, $6M_p/l$
5. 3 m   7. $4M_p/l$, $4M_p/l$, $3M_p/l$, $2M_p/l$, $1.333M_p/l$
8. 1.5, 2 9. 85 kN m, 1.046
10. 1.651, 40.45 kN m, 1.633 11. 1.92, 1.194
12. 1.481, 1.591 13. $(3 - 2\sqrt{2})Wl$

Chapter 4

1. 1.573 2. 1.679, 1.459 3. 1.524
4. 1.382 5. 1.667, 1.756 6. 28.13 kN m, 81.2 kN
7. 1.65, 1.5, 1.611 8. 1.515, 1.449
9. 1.556 11. $3M_p/R$, $8M_p/R$ 12. 1.5

Chapter 5

1. $(12ln3 + 8\sqrt{3} - 10)\delta_0/15$ 2. $3.2M_p l^2/6EI$, $2.8M_p l^2/6EI$
3. $0.537M_p l^2/EI$ 4. $10M_p l^2/9EI$
5. $13M_p l^2/12EI$, $4M_p l^2/27EI$; $5M_p l^2/6EI$, $5M_p l^2/18EI$
6. $1.36M_p l^2/EI$

Chapter 6

1. 52.14, 42.49 kN m
2. $0.12a^3 \sigma_0$; $M_N = M_p(1 - n^2/3)$ for neutral axis in flange,
   $$M_N = M_p(1 - 5n^2/3) \text{ for neutral axis in web}$$
3. $M_N = M_p \cos n\pi/2$
5. 421.5, 391.2 kN

Chapter 7

1. 31.67, 25 kN m   2. $\beta_1 = \beta_2 = 46.67$ kN m
3. $\beta_1 = \beta_2 = 56$ kN m

Chapter 8

2. 1.333, 1.5   3. 1.364, 1.5   4. 1.6, 1.6
5. 1.35, 1.5   6. $7.322M_p/l$, $8M_p/l$ 7. 1.371
8. 1.481   9. $1.829M_p/l$ 10. $0.138M_p l^2/EI$

Indexes

Author Index

Subject Index